The Agri-Environment

The application of ecological theory and conservation biology to agricultural ecosystems has become an important and growing research field and undergraduate course component in recent years. This book is both an academic textbook and a practical guide to farm conservation, and has evolved from the authors' extensive teaching experience. It covers the ecology of farmed land, how agricultural practices influence the environment, how agriculture has changed over time and how the species that inhabit the agri-environment have adapted. It also covers the history of agricultural policy and subsidies and the development of agri-environment schemes. A number of different farming systems are discussed, as are the difficulties in determining their relative merits. Guidance is offered on how to produce a workable farm conservation plan, and the final chapters look to the future and the development of new greener farming systems.

JOHN WARREN has 20 years of research and teaching experience in agro ecology, at the University of Liverpool and the University of the West Indies. He spent 7 years at the Scottish Agricultural College as head of the Farmland Ecology Unit, and is currently Director of Education and Lecturer at the Institute of Rural Sciences, University of Wales, Aberystwyth.

CLARE LAWSON has 10 years of research experience in habitat restoration and ecosystem processes within the farmed landscape, working at the University of London, Cranfield University and latterly as a Research Fellow at the University of Reading in the Centre for Agri-Environmental Research (CAER). Her current research focuses on the impacts of management practices on farmland biodiversity and the importance of trophic interactions (plant-soil and plant-insect interactions) in grassland restoration.

KEN BELCHER is an Associate Professor in the Department of Agricultural Economics at the University of Saskatchewan, Canada. Prior to this, he was a waterfowl and wetland biologist and he maintains a strong interest in the interactions and potential conflicts between agricultural systems and wildlife conservation.

The Agri-Environment

JOHN WARREN
University of Wales, Aberystwyth, UK

CLARE LAWSON
University of Reading, UK

KENNETH BELCHER
University of Saskatchewan, Canada

CAMBRIDGE
UNIVERSITY PRESS

Shaftesbury Road, Cambridge CB2 8EA, United Kingdom

One Liberty Plaza, 20th Floor, New York, NY 10006, USA

477 Williamstown Road, Port Melbourne, VIC 3207, Australia

314–321, 3rd Floor, Plot 3, Splendor Forum, Jasola District Centre, New Delhi – 110025, India

103 Penang Road, #05–06/07, Visioncrest Commercial, Singapore 238467

Cambridge University Press is part of Cambridge University Press & Assessment, a department of the University of Cambridge.

We share the University's mission to contribute to society through the pursuit of education, learning and research at the highest international levels of excellence.

www.cambridge.org
Information on this title: www.cambridge.org/9780521614887

© J. Warren, C. Lawson and K. Belcher 2008

First published 2008

A catalogue record for this publication is available from the British Library

ISBN   978-0-521-84965-4   Hardback
ISBN   978-0-521-61488-7   Paperback

Contents

Preface and Acknowledgements

Historically agriculture has been considered purely as the method by which humans produced most of their food, fibres and other natural products. This activity has dramatically altered the farmed environment, favouring some species and habitats and degrading or destroying others. Over time as the human population has grown and agricultural activity has intensified the magnitude of this effect has increased, resulting in recent rapid declines in the abundance of many species and in the conversion of semi-natural habitats to monocultures. Population crashes in many of the species associated with farmed land and reductions in the quality and quantity of ecological services delivered by farmed land have resulted in an awareness that agriculture produces much more than just food. This realisation combined with other pressures such as reducing and decoupling economic subsidies from food production and changes in consumer demands for ecological goods and services are driving a second truly green revolution within the agricultural industry. New understanding of the ecology of the impacts of agriculture at a range of levels is enabling multifunctional production systems to be designed that deliver quality food products while supporting biodiversity and maintaining ecological services. Everyone involved in the agricultural industry during the twenty-first century will need an understanding of how to balance these conflicting demands.

This book has been written for agricultural and conservation students and researchers and for those actively involved in balancing food production with on-farm conservation. Its aims are to provide an understanding of the underpinning ecological science that regulates the plant and animal populations and communities that inhabit the agri-environment. Through these ecological processes the human activity of food production changes the environment in which we co-inhabit with the other species on the planet. It is therefore essential that we understand these mechanisms if we are to better manage them in future. But agriculture is not purely an ecological science, it also has social and economic elements and this book covers the history of agricultural policy and subsidies and the development of agri-environment schemes. A number of different production systems (some more scientific than others) are available which, at least in part, attempt to balance agricultural production with sustainable environmental management. These alternative production systems are explored, as are the difficulties in determining their relative merits. For the moment the main policy mechanism used by western governments to encourage more ecological sustainable farming is the agri-environment scheme. The principles behind such agri-environment schemes are discussed and a guide is provided for how to produce a workable farm conservation plan. The final chapters cover recent developments in our understanding of the importance of scale and landscape complexity within the agri-environment. These concepts are becoming increasingly important in managing farmed landscapes, for example in locating habitat restoration projects and increasing habitat connectivity. Such factors will be important from the level of farm planning to designing national policy. It is always dangerous to predict the future, but balancing the partitioning of sunlight energy as food between an increasing human population and the other species that share our planet is a difficult problem and one that requires a great deal of scientific understanding.

We wish to thank many of our colleagues and friends for discussions on the various chapters, for commenting on parts of the text, or for simply enduring our trials and tribulations. Members of staff of the Institute of Rural Sciences at Aberystwyth University, Sue Fowler, Will Haresign, Graham Harris and Mike Rose have rendered valuable assistance, as have Chris Topping and Bryony Williams. To all who have contributed in any way we wish to express our deepest appreciation.

1

An introduction to agro-ecology

Introduction

Agriculture, the cultivation of plants and domestication of animals by humans, is approximately 10 000 years old. In evolutionary terms this should be an insignificantly short period of time, but it has not been. Human agricultural activity has changed the world completely; the genotypes of domesticated species have often changed beyond recognition. The relative abundances of species on earth have been altered dramatically, so that previously uncommon weedy grasses (cereals) now dominate vast areas. Even the habitats occupied by wild species have frequently been modified so they now support entirely novel communities of plants and animals. Natural communities from late in succession have been replaced by communities with ecologies more typical of early succession. The move from hunter-gathering to farming has allowed the human population to rise to more than six billion and therefore everything that humans do, every impact that we make on the planet, can be considered as an indirect environmental impact of agriculture. However, the scope of this book is less ambitious as it covers the more immediate direct interactions between agriculture and the environment. The function of agriculture is to direct energy from the sun (including fossil sunlight) into the human food chain. Little of this energy that is utilised by humans is then available for the other inhabitants of our planet. This process involves a great deal of effort to convert natural habitats into agricultural ones and replace wild species with domesticated ones, while natural ecological processes are exerting pressure on the system in the opposite direction. This movement away from the natural situation constitutes one of the direct environmental impacts of agriculture.

This book explores the nature of these impacts, how they can be managed, and whether they can be balanced by farmers and policy makers with our need to produce food. To better understand the complexities of the environmental impacts of agriculture, this first chapter explores the origins and ecologies of species that inhabit farmed land. This understanding of the population ecology of single species is developed into looking at competitive interactions between species, which builds into community ecology theory. Finally an understanding of the management and exploitation of biodiversity within the agricultural context are introduced as key themes that are considered further throughout the rest of the book.

Species that inhabit farmed land

Farmed land has only existed for, at most, about 10 000 years, which is very little time for new species to have evolved which are adapted to this relatively new habitat. During this period, selection by humans has produced a range of domesticated crops and animals that are no longer able to survive without the assistance of humans outside the agro-ecosystem. They have been so genetically modified, by hybridisation and selection, that their origins were uncertain until the advent of modern molecular genetics (Hancock, 2005). Although less dramatic, the wild species that co-inhabit farmed land have also undergone sometimes substantial genetic changes. Many of the wild species of the arable agri-environment would have been rare or out of range before the advent of agricultural activity. These plants and animals had evolved in naturally disturbed habitats associated with early succession, such as sand-dunes, retreating glaciers or volcanic lava fields. Such species are known to ecologists as ruderals, they have life-histories characterised by short lifespans, the production of large numbers of small offspring, and they are highly mobile and invest few resources in defence mechanisms. These are the annual weeds of the plant world. Many agricultural invertebrate pests and diseases have similar life-history strategies. Alternatively these problem species can be seen as valuable biodiversity at the base of the food chain for the other larger more charismatic species of farmed land. Determining to what extent we tolerate these non-agricultural species diverting sunlight from the human food chain to the rest of nature is central to how we manage the agri-environment and this is a problem that we will return to again in Chapters 8 and 10 (Figure 1.1).

Until recently pastoral agriculture was based on the grazing of native or semi-natural grasslands or dwarf-shrub communities, but the twentieth century saw an increasing reliance on more productive, agriculturally improved forage systems. These new artificial grass-dominated communities are species-poor. The species they contain are now amongst the commonest on Earth, but their natural ecologies and genetics have been completely changed by agriculture

Figure 1.1 When should a species be considered an agricultural pest or be tolerated or even encouraged as valuable biodiversity? The answer to this question is central to how we think about and manage the agri-environment.

(Warren *et al.*, 1998). The plant species that form the basis of both improved and semi-natural pasture systems evolved in non-agricultural grasslands. In Western Europe most grasslands have previously been thought of as transitionary vegetation communities, which form part of a succession that would naturally lead to climax woodland. Vegetation succession has been arrested at the grassland stage only because of agricultural grazing. This view has been challenged by Vera (2000) and many now think that the natural vegetation of Western Europe may have included much more grassland than was previously considered. This is significant because it might imply that agricultural habitats regarded as semi-natural may be more natural than previously thought and the species associated with them may have been coevolving for longer. Where the history of agricultural development is much shorter, such as North America and Australia, there is a better understanding of the make-up of the climax communities, whether forest or grassland. In fact remnant tracts of many natural vegetation communities, albeit very small in some cases, do still exist.

Population dynamics of single species

The science of ecology is about understanding why species live where they do and why sometimes they are abundant and sometimes rare. The practice of agriculture is about managing populations of species so that they can be exploited by humans. Therefore, by necessity agriculturalists need to know what species will live where and how well they will thrive. Agriculturalists need to understand ecology and need to know what regulates populations. Understanding what processes regulate population sizes underpins selecting stocking rates of livestock, sowing rates of crops, what species can be successfully grown or kept together, plus the biological or chemical control of pests and diseases.

With a single species, in the simplest of all worlds, that is with no overlap of generations, no immigration or emigration from the population and all individuals being hermaphrodite, all of whom successfully reproduce, because resources (food, water, space, sex, etc.) are in excess and disease, predators and competitors are all absent, then:

$$N_{t+1} = N_t R.$$

The population in the next time period (N_{t+1}) = the population now (N_t) multiplied by the maximum number of offspring an individual can produce, R.

These restrictions might seem unrealistically crude; however, the population dynamics of many species of weed and pest of agriculture can at least spasmodically be regulated and mathematically predicted by such exponential explosions in numbers when they exploit a new resource, for example a newly ploughed field. These ruderal species tend to have populations that rapidly increase in numbers and then crash, with the episodic declines usually resulting from agricultural activity, such as ploughing or the application of chemical control.

In most species, the size of the population is regulated by density-dependent processes. That is, as the population size increases competition between individuals of the same species tends to reduce the growth rate of individuals, which affect the age or size at which individuals reproduce, decrease the birth-rate or increase the death-rate. Exactly what combination of these possible effects occurs differs between species, but the outcome of limiting population size always arises. This within species competition for resources, which reduces the size of individuals and over time increases the death-rate of smaller (less competitive) individuals, is responsible for a relationship known as self-thinning in plants (see Figure 1.2) and this effect is behind what determines optimal sowing rates for crops and planting densities for tree crops.

Even if agriculturalists are not consciously aware of the self-thinning rule, they select sowing rates for crops so that the plants are able to grow to a desirable size by keeping levels of intraspecific competition low enough to avoid crop plant mortality. This must be balanced by sowing enough of the crop to obtain an acceptable yield and for interspecific competition to be intense enough to help in suppressing the growth of non-crop plants.

In managed agricultural populations extra resources are used to counter the effects of density dependence to artificially increase birth-rates. Death in

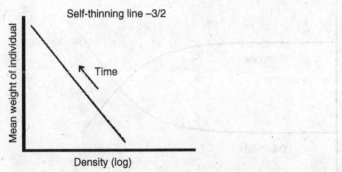

Figure 1.2 As individuals grow over time competition becomes more intense. This results in the death of the weaker/smaller individuals, which reduces the density of the surviving population, which eases competition and allows the surviving individuals to grow larger. Thus both size and population density are interrelated and change over time in accordance with the self-thinning rule. The gradient of the self-thinning relationship −3/2 arises from the fact that density (log) is area based and changes as a square whereas weight/volume changes as a cube.

domesticated species tends to escape density dependence by being regulated by harvesting/slaughtering rather than competition. However, the natural processes illustrated in Figure 1.3 do regulate the populations of the wild species that inhabit the agri-environment.

Mathematically, density dependence can be incorporated in population equations, with similar assumptions as before, those of: no overlap of generations, no immigration or emigration from the population and all individuals being hermaphrodite; although competition within a species is represented, the effects of disease, predators and other competitors are again all absent. Under these conditions:

$$N_{t+1} = \frac{N_t R}{(1 + aN_t)^b}$$

As before N_{t+1} represents the population size in the next time period, N_t is the population now and R is the maximum number of offspring an individual can produce. The only new parameters in the density dependence equation are a, which is described by some plant ecologists as 'the area of isolation' (that is the area which a plant needs to be able to produce R seeds and beyond which no extra seeds are produced) and b, 'the coefficient of resources use efficiency'. However, both these values are probably best thought of as simply constants, which just happen to be useful in predicting the size of the population next year. The effect of variation in the value of parameter b on the population size in the following time period can be seen in Figure 1.4. Species with low b and R values and hence relatively stable populations are associated with late succession, such

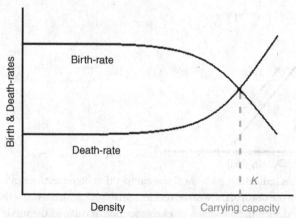

Figure 1.3 In the wild as population density increases, birth-rate decreases and death-rate increases. At the point at which the birth-rate and death-rates are equal, recruitment and death are equal and the population size may reach a stable equilibrium size. This is known as K, the carrying capacity.

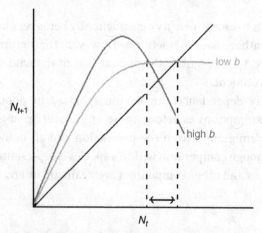

Figure 1.4 Variation in the value of parameter b affects the robustness of the predicted population size in the next time period (N_{t+1}). When b is low the predicted population curve cuts the 45° line close to the horizontal, so that a small amount of variation in the current population (N_t) has very little effect on the prediction of the subsequent population. However, when the value of b is large the predicted population curve cuts the 45° line in such a way that a small level of variation in the estimated population now makes a great difference to the predicted size of the next (future?) generation.

as oak trees or large mammals; those with high values of b and R, which are prone to dramatic changes in population size, are more likely to be associated with agriculture, such as locusts.

Species that are pests of agricultural systems tend to have the capacity to produce large numbers of offspring (they have large values of R) and therefore

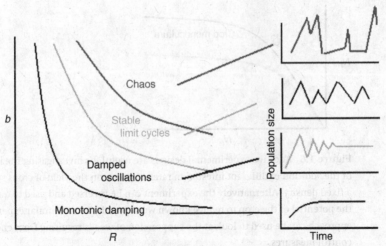

Figure 1.5 Knowing the parameters R and b allows long-term predictions to be made about the stability of population size. Species that are pests of agricultural systems tend to have populations that have chaotic dynamics and are prone to rapid increases and decreases that are difficult to predict.

their populations have the ability to increase very rapidly. When this is combined with high values of b, which make it difficult to make reliable predictions of the population from generation to generation, then the long-term population dynamics of agricultural pest species can be difficult to predict (see Figure 1.5). However, the chaotic population dynamics of many agricultural pests does not mean that their populations cannot be predicted, just that increasing amounts of data are required to successfully predict over reduced periods of time. Plus, given that many of the apparently random population crashes result from agricultural control measures, it is not true to say they are genuinely chaotic.

Two species interactions in agriculture

Much of the above discussion of the population dynamics of single species considered pest species, but of course these do not live as single species, and although intensive agriculture is often regarded as the management of monocultures, the reality is rarely so simple. In many farming systems managing different species together in the same space at the same time is the norm; therefore, if we are to successfully control pests or optimise yields over several species, we need to develop our understanding of the population ecology to more complex systems.

Two different experimental approaches have been developed by crop-ecologists to investigate the competitive interactions between two species. The two methods

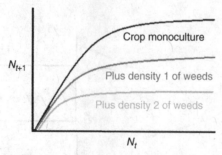

Figure 1.6 Additive experimental designs are useful for investigating the impact of the addition of different infestation rates of weeds on the yield of crops sown at a fixed density. Alternatively the experiment can be reversed and used to quantify the potential of the crop to reduce known weed populations. A similar experimental approach can be used to look at the suppression of weeds resulting from chemical control measures.

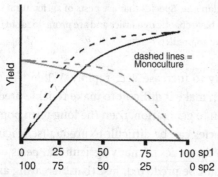

Figure 1.7 Replacement experimental designs have a fixed sowing density, but within a plot the ratio of two species varies from monoculture of one species through to the monoculture of the second.

relate to different applications. Firstly, additive experiments (see Figure 1.6) involve the addition of different levels of a second species to a fixed population of the first species. This can be used to represent the occurrence of a population of weeds or a second crop species in a fixed sown population of a crop. Secondly, there are replacement series experiments (sometimes called De Wit replacement experiments in honour of the Dutch ecologist who developed the approach), in which individuals of one species are replaced by individuals of a second, but with the overall population being kept constant (see Figure 1.7). This second approach is useful when trying to establish the optimal ratio of two species to use when bi-cropping.

Replacement experiments typically demonstrate a phenomenon of fundamental significance to agro-ecology. Competition between species is usually less intense than is competition within a species. This is because individuals of the same species have the same environmental requirements, they compete for exactly the same resources. In contrast, different species will have different resource requirements, they will need different nutrients, or may root at different depths or grow at different times of the year, etc. Two very important facts result from this:

1. Overall yields (in terms of biomass production) tend to be higher with two species than in monoculture.
2. Such species have the ability to coexist by competing for different resources and so diversity is assembled.

Of course reality is more complex than this simple assertion, but it is an important factor that operates behind many agricultural processes. The first complication arises from the experiment's simple assumption of a fixed ratio of species. Just because two species are sown at a fixed ratio does not mean that they remain at that ratio; this is particularly true where there is differential growth or spread, such as with vegetative species. Secondly, in the artificial situation of a replacement experiment both species are usually established at exactly the same time. In the field, however, species may establish at different times or over a period of time. This can be important in further promoting diversity, because species that are competitively inferior (and over time would be lost due to competition) may not be excluded by competition if they have the opportunity to establish before the normally competitively dominant species. In addition, from an agricultural point of view the simple statement that overall yields are higher with two species than one may have little value, because the yield of total biomass may be less useable and there can be many practical problems in the management and harvesting of more than one species. Certain varieties of cereals and legumes can successfully be combined together and their grains separated mechanically, but other combinations with different maturation times can be more difficult to process.

An additional complication in the agricultural application of replacement experiments is that the outcome is often density dependent and such experiments are typically carried out at a single fixed sowing density. When a replacement experiment is performed at low density there is plenty of opportunity for the two species to exploit different resources (and therefore have higher yield in comparison to monoculture). However, when the same experiment is repeated at a higher overall sowing rate, the level of competition between individuals is more intense and the subtleties of between species differences are reduced as individuals struggle to survive, so that the increased yield potential of bi-cropping is reduced. Therefore, if replacement experiments are to be used to

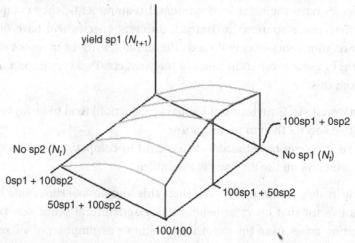

yield sp1 (N_{t+1})

100sp1 + 0sp2

No sp2 (N_t)

No sp1 (N_t)

0sp1 + 100sp2

100sp1 + 50sp2

50sp1 + 100sp2

100/100

Figure 1.8 Response surface analysis of competition between two species over a range of ratios and overall sowing densities can be used to optimise bi-cropping systems or to identify optimal stocking rates and ratios of different livestock species such as cattle and sheep. In this figure the x and y axes are N_t and N_{t+1} as in Figure 1.4 while the z axis (N_{t2}) represents the size of the sown population of the second species at the time 0.

optimise ratios and sowing densities in bi-cropping systems, a series of experiments is needed, over a range of sowing densities (see Figure 1.8).

Parasites, pests and diseases

So far we have been considering the population dynamics of two competing species within agricultural systems such as weeds and crops, two species of grazing animals or combining two crops. This situation is different when one of the species is a domesticated species and the second is a direct predator, parasite or disease. Above we saw that many pests and diseases have the ability to rapidly increase in numbers to exploit available agricultural resources; their large reproductive potential allows them to generate lots of viable offspring, which are the raw material upon which natural selection acts in adaptive evolution. The rate of evolution of agricultural pests can be rapid for two reasons: firstly the large numbers of individuals involved, this does not just reflect the large numbers of progeny produced but also the vast areas of agricultural production over which they are produced; secondly the intensity of the selection applied by chemical, biological, genetic or physical means can be intense. It is no surprise, therefore, that when agriculturalists try to produce enough food to feed a global human population of six billion plus, other species adapt to exploit this vast potential food resource. Whatever control measures are applied, pests seem to evolve mechanisms of resistance, be they insecticide, herbicide or fungicide resistance in crops, or antibiotic or anthelmintic resistance in livestock, plus the ability to break down genetic resistance in both.

The rapid evolution of agricultural pests and diseases is an example of an evolutionary phenomenon know as the Red Queen Effect (Van Valen, 1973). What is occurring in these situations is an evolutionary arms race between the breeder or chemist and the pest or disease. Each time a breeder produces a new variety or breed with a resistance gene or a chemist produces a new agrochemical or veterinary medicine it imposes a selection pressure on the pest to evolve a mechanism to avoid the method of control. Once the pest has evolved its own resistance, then the new variety or chemical control becomes ineffective and the breeder and chemist are ensured of employment as they are required to develop a new form of control. Some agrochemicals or resistance genes may be more difficult than others for pests to evolve resistance to, but given time they will. This coevolutionary process occurs in nature, driving arms races of defence mechanisms and counter-mechanisms in pests and diseases and their hosts. For this reason abundant species are unlikely to reproduce vegetatively for too long before they become too badly infested with pests and disease. Similarly, all agricultural crop varieties and agrochemicals will have relatively short periods of effectiveness. The more widely used they are, the stronger the selection pressure they will produce, and the shorter their shelf-life is likely to be.

From a profitability perspective, an ideal new pesticide is one in which resistance naturally evolves in the pest population at around the time the patent on the product runs out. This strategy prevents commercial competitors from being able to exploit an innovative company's research and development costs. While this might make good economic sense in the market economy, it is not a sustainable way to manage resistance genes, antibiotics or agrochemicals. Away from market economics, there is a method to escape from this evolutionary treadmill in the managed agricultural environment. Red Queen evolutionary arms races are linear in nature. Evolution in agricultural pests tracks genetic changes that occur in their host or is driven by a new control method until such a point that resistance genes spread throughout the pest population. However, if the selection pressure applied by the new crop resistance gene or chemical control agent was varied in space or better still in time, then the strength of the selection pressure would be reduced or completely altered in direction. Utilising different resistance genes in different locations is part of the rationale for growing different cultivars in adjacent fields or more rarely as multi-lines mixed within a single field. This has the advantage that the crop is less likely to be devastated by a particularly virulent strain of pathogen or pest. However, this method of managing the evolutionary arms race that occurs between agricultural hosts and their pests merely slows down the pace of genetic change. Exploiting different resistance genes or varying chemical control methods over time, rather than space, totally changes the nature of evolution. Instead of tracking the evolution of its host the pest species is required to evolve in a different direction every time the cultivar or agrochemical etc. is changed. This of course requires large-scale coordination of the industry and requires companies to take their products or varieties off the market for a number of years, and therefore it is unlikely to be compatible with free market economics, but it would enable a more sustainable way to manage pest control in the agri-environment. This approach has been successfully applied in nature. Two very different groups of species have effectively evolved this method of avoiding their pests, by synchronising their life-cycles and being unavailable as a food source for several years. A further refinement to this strategy is the use of prime numbers, so that when the food resources become available, it is difficult to predict exactly when they will appear. The species involved are cicadas and bamboos. Certain species of cicadas emerge as adults after 7, 13 or 17 years as larvae. In the intervening years no adults emerge, so this food source is unavailable for their pest and disease species to attack. Similarly, bamboos synchronise their life-cycles, with all individuals within a species flowering and setting seeds in the same year before dying. When this mass production of seed occurs a huge food source is produced, but potential consumers are unable to predict the timing of the event, as in some species it occurs only every 120 years. Such a long-term removal of a resource is not practical within agriculture, but crop rotations (which also tend to be based around prime numbers) have similar if less dramatic effects. However, if a particular crop could be removed from cultivation for more than a hundred years, it would probably be freed from many of its pest species by the process. Similarly, if a pesticide or antibiotic could be withdrawn for such a long period, there would be few resistant individuals left to pass on their resistance genes when usage was resumed.

It is well known that in natural systems predators and prey or diseases/parasites and their hosts tend to regulate the size of each other's population via a mechanism known as predator–prey cycles. As predators or disease-causing species increase in abundance they reduce the population of their prey or host species, reducing their own food supply until the population of predators declines to such a point that the prey population is able to recover. Such natural regulation of populations is often spoken about by organic agri-culturists, but it is difficult to find predator–prey cycles operating in most agricultural systems for three reasons. Firstly, pests of agricultural systems tend to be generalist species; because their ecological interactions are young in evolutionary terms, pest species typically have the ability to attack a range of hosts so that when the host population declines, rather than track this decline, the pest species moves over to an alternative host. Secondly, the large-scale production of agricultural species enables pest species to potentially attain very high numbers, because if their host population declines in one area, the pest species is likely to have the ability to relocate to a neighbouring farm. Under these outbreak conditions the third process is likely to operate, and human intervention via chemical or mechanical control is likely to be applied to the pest species. In small-scale organic systems predator–prey cycles can be effective, because the second of these processes does not apply, and it may function further up the non-agricultural food chain. Crop rotations can be highly effective in regulating pest populations, but when production is scaled up to the industrial level of modern production farming pest populations can quickly move from host to host (see Chapter 9). Separating production into small blocks may alleviate this difficulty, but large-scale agricultural production will always produce a large potential food source for pest species to exploit. Fighting this never-ending ecological and evolutionary battle, while still allowing sufficient photosynthate to enter the non-human food chain, is the main challenge that faces modern agriculture.

Biological control and chemical control

The biological control of agricultural pests in its simplest form is the use of one species to control a second species, and as such it depends on the two species population dynamics described above. Classically, predators or diseases have been released to reduce the numbers of agricultural pests. Biological control has often been regarded as an alternative to the use of chemicals, but there is no reason that the two methods cannot be complementary. Although the use of chemicals to control pests in agriculture can be traced to 4500 BP when the Sumerians used sulphur compounds to control insects, and later the ancient Chinese used plant-derived complex organic insecticides, their intensive use was a twentieth-century invention. The widespread use of chemical pesticides in agriculture has tended to be characterised in the literature as being environmentally damaging; in contrast, biological control has been seen as being natural and environmentally benign. However, conservation ecologists are slowly starting to realise the damage that has been inflicted by poorly considered attempts to use biological control (Hamilton, 2000). In fact there are many parallels between the development and environmental impacts of these two different control methods.

The first generation of synthetic pesticides that were widely used were developed during the Second World War in an attempt to eradicate malaria mosquitoes. The insecticides, DDT, chlorinated hydrocarbons, organophos-phates, carbamates, and herbicides, 2,4-D, DNOC, MCPA, were broad-acting and toxic to a wide range of different species (see Chapter 3). These chemicals are now known to have had several undesirable environmental effects includ-ing bio-magnification, killing non-target species and having long half-lives. Subsequent generations of pesticides have tended to be better targeted both in their chemical specificity and by refinements in the designs of spraying equip-ment. Many agrochemicals are now highly complex organic molecules, which are applied in low doses and rapidly break down in the environment. Their direct environmental impacts in terms of poisoning of wildlife are considerably less than those of the first generation of pesticides. However, any efficient method of pest control is likely to have significant impacts for species further up the food chain.

The first attempts to use biological control were also unrefined and resulted in unexpected ecological impacts. Classical biological control involves the release of predators or diseases, typically to control an introduced alien agricultural pest. The problems with this approach have been that the introduced control species frequently fails to establish, and when it does it may unexpectedly attack native species, driving them to extinction. Furthermore, classical biological control that affects non-target species has been associated with the 'genie out of a bottle' problem that has been levelled at the release of genetically modified organisms, in that once a biological control agent has been released into the wild, it can itself be difficult to control, if it starts to behave in an unexpected way. Perhaps the best known example of this is *Bufo marinus*, the cane toad that was introduced into Queensland in 1935 in an attempt to control cane beetles. Since then it has spread west and south across Australia, eating or poisoning much of the native wildlife. There are many such examples mostly involving insects, but arguably the most significant in terms of causing extinctions has been the introduction of the predatory snail *Euglandina rosea* from the United States with the intention of controlling the giant African snail *Achatina falica* that was widely introduced across Asia and the Pacific as food. Unfortunately this introduction has resulted in the decline and extinction of many endemic snails of the genera *Achatinella* and *Partula*. Over time biological control measures have also become more refined. A whole range of techniques are now covered by the term biological control, including the augmentation of wild populations of natural enemies, or enhan-cing these natural populations by habitat management (e.g. using beetle-banks) or inoculation of these naturally occurring populations by the periodic or one-off releases of individuals. All of these techniques are more targeted than classical biological control and since they all avoid introducing alien species, they are free from the 'genie out of a bottle' problem.

Multiple species interactions and communities

So far we have just considered the ecological interactions that occur between pairs of species within the agri-environment. These interactions are relatively simple to represent mathematically by extending the single species equations to include two species, by numerically converting one species into another using a coefficient that represents their relative competitive abilities. Although this approach has been extended to three and occasionally more species it rapidly becomes unworkable, because the competitive relationship between two species may be changed by the presence of a third. Therefore an entirely different approach is needed if we are to understand the processes at work in agricultural communities comprised of several species.

When several species regularly occur together as they do in agricultural grasslands, then it is not simply a case that similar grasslands contain similar lists of species, but a few species are predictably always common while most species are typically present in much lower numbers. The extent of this varies between grasslands. In agriculturally improved grasslands, which are species poor, the few species present are highly abundant; in contrast, in old semi-natural grasslands most species present occur in low abundance with only a few more common. So what are the ecological mechanisms at work which regulate species richness and abundance within permanent agricultural communities? Opinions are divided on this; there are two main theories both of which may be correct in different cases and both have interesting implications for agricultural management. The most widely held theory is called the complementarity theory and it contends that potentially all the species present within a community contribute to community functioning. This is supported by several experimental studies, which appear to show that the addition of more species to a plant community increases the community's ability to produce biomass and to recover following environmental perturbation such as drought (Tilman *et al.*, 2001; Steiner *et al.*, 2006). The alternative theory, called the selectionist theory, maintains that community functions are determined by the few abundant or dominant species within the community. Supporters of this theory point out that although complementarity predicts that the highest levels of biomass production should be produced from species-rich communities, in fact the most productive of all grasslands are agriculturally fertilised species-poor pastures (see Figure 1.9). Furthermore, only the dominant species are reliably present within a community to ensure it continues to function. In many cases the less abundant species are absent from a community, but still the community is recognisable and functions no differently for the absence of a few species.

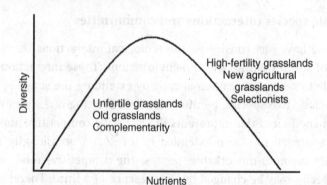

Figure 1.9 The relationship between species diversity and fertility in grassland communities is humpbacked. Under very low nutrient conditions, such as pioneer sand-dunes, very few species can survive. As fertility increases so does species richness, with the highest diversity levels being associated with old semi-natural meadows. These species-rich pastures have had a long evolutionary history providing plenty of time for complex ecological interactions, of the type required for complementarity theory to function, to develop. Under the artificially high levels of nutrients associated with modern intensive agricultural grasslands, very few species dominate the swards. These grasslands have no great evolutionary history and it is unlikely that species present will interact in a complementary fashion as they find themselves in a new assemblage of species in a new environment.

Very different implications emerge from these two contrasting theories for both agricultural production and biodiversity conservation. If complementarity theory is correct, then potentially increased agricultural yields can be achieved by exploiting different species' ability to utilise different niches within the pasture. Under these conditions it is important that all species within the community are conserved if we wish to protect community functions. In contrast if a few dominant species are responsible for the vast majority of the biomass production, and the presence of additional species does little for community productivity or stability, then the future of agricultural production lies in monocultures and low diversity, and the rationale for conservation must be ethical, or long-term, rather than current utility. This is perhaps not a simple comparison, because complementarity of niche exploitation by different species must take a long time to develop, whereas artificially high-fertility pastures are a development of modern agriculture. It is perhaps not surprising therefore that selectionist mechanisms appear to operate in new intensive agricultural grasslands whereas complementarity functions in old species-rich pastures (see Figure 1.9). Agricultural scientists have spent a considerable amount of time and effort as geneticists, agronomists, engineers and chemists trying to improve the yields of species-poor systems, and as yet virtually none on exploiting complementarity and developing the yield potential of more species-rich systems. Only in the tropics is polyculture widely practised, and then typically not in large-scale production systems, as it tends to be highly dependent on cheap human labour to harvest different crops at different times.

Semi-natural habitats and agricultural management

So far the above discussion of the ecological mechanisms that determine the composition of agricultural plant communities has focused on internal processes that emerge from the species themselves. However, abiotic edaphic factors such as soil chemistry and climate, as well as agricultural management, all combine to determine which species are able to coexist and their relative abundances within permanent pastures. In old pastures that have never or at least not recently been agriculturally improved by combinations of fertiliser applications, liming, draining and reseeding, acid grasslands tend to be less diverse than neutral grasslands, which in turn are less diverse than calcareous grasslands (Table 1.1). So care needs to be taken when using the term species-rich grassland to ensure that this is relative to the potential for the soil type.

When a particular agricultural management is applied to a pasture under specific environmental conditions for long enough and given adequate seed supply, then a recognisable and predictable plant community will develop. Because soil types and agricultural management tend to fall into discrete classes, then the plant communities associated with them are also generally recognisably discrete, although some vegetation types are more distinct than others. For example, the ecological discreteness of a salt-marsh community tends to be more sharply defined (by the tidal influx of seawater) than are grassland communities determined by soil fertility. In spite of this problem ecologists around the world have developed methodologies for recognising and describing

Table 1.1. *The term 'species-rich' when applied to grassland communities must be relative to the soil type as calcareous grasslands have the potential to be much more diverse than do neutral or acid grasslands*

Type of grassland	Species richness
Calcareous grassland	30 sp/m^2
Neutral grassland	20 sp/m^2
Acid grassland (calcifugous grasslands)	10 sp/m^2

plant communities, many of which can be associated with particular types of agricultural management. Some of the plant communities associated with agricultural activity, for example hay meadows, are recognised at least loosely by the public and can be said to have cultural significance. Since many of the traditional agricultural management practices that have given rise to these communities are now regarded as redundant, their associated communities have become increasingly rare (see Chapter 3). As a result some of these now rare agricultural plant communities may be recognised by conservation legislation or farmers may receive payments for maintaining them. These communities are, however, semi-natural and to an extent arbitrary, in that their botanical composition is determined by human activity and if a different form of agriculture had developed we would now know a different set of vegetation types. Paying farmers to carry out or reinstate redundant agricultural practices in order to maintain arbitrary plant communities must be regarded as wildlife gardening. This may not be a bad thing, but it may not be sustainable in the longer term. The challenge remains whether we can exploit the ecological principles described above and develop a form of agriculture that is productive and compatible with maintaining diversity.

Summary

Sometimes knowingly but often unwittingly agriculturalists use ecological principles in regulating population size and yields of domesticated species. Agriculture is applied ecology; it manipulates birth-rates and death-rates of single species by controlling population sizes or by avoiding density-dependent processes by supplying additional resources. Competition between species is avoided when trying to optimise yields in multiple species farming systems. The effects of competition between species are also what the farmer has to manage when trying to prevent yield losses caused by pests and diseases. This is a never ending ecological/evolutionary struggle for the agriculturalist. As we shall see in subsequent chapters, in the recent past, mankind has become very efficient in fighting this battle and at diverting photosynthate into the expanding human food chain, which has had dramatic impacts on the rest of nature. Much of the rest of this book investigates how agriculture can balance feeding a vast human population whilst maintaining a diversity of other living things together in a healthy environment for all these species.

2

Agricultural support and environmentalism

Introduction

This chapter outlines some of the key events in agricultural policy from the beginning of the twentieth century until the recent reforms of the European Common Agricultural Policy (CAP) in 2003, the 2002 US Farm Bill and ongoing world trade negotiations. It describes the widespread introduction of subsidies to support farm prices and the unprecedented expansion of agricultural production, to the advent of food surpluses and concerns over the environmental impact of modern agricultural practices. The development of agri-environment measures and the change in emphasis from an agricultural policy that supports production agriculture to one that supports the environment and rural development is explained and the principles behind agri-environment measures examined.

Agricultural policy: the start of government intervention

The regulation of agricultural markets and intervention by national governments to support farm incomes is not a new phenomenon. Throughout the course of history national governments employed various policies to support and protect agricultural production, such as the Corn Laws designed to protect British cereal farmers from foreign imports. However, it was not until the long-lasting economic depression of the 1920s and 1930s, which was also a period of agricultural depression with low commodity prices and depressed farm incomes, that national governments systematically intervened in agricultural markets to ensure the home production of food and to support their national industries. In the United States, following increasing political support for the principle of market regulation, President Roosevelt introduced a system of price supports and incentives in 1933 as part of the 'New Deal'. Across Europe, a range of different measures were introduced by national governments to protect farm incomes, including levies on imports and price support for certain agricultural products. In the United Kingdom, the Royal Commission on Agriculture in 1928 advised that areas of potentially good land should be brought into production, in order to provide sufficient food. This was followed during the 1930s by the government's introduction of measures to support the cereal and dairy sectors, with the introduction of deficiency payments for cereals and the establishment of the Milk Marketing Board.

The impact of the war in 1939 on agriculture was immediate. In the United States, agricultural production and farm prices increased as the demand for food rose as European agricultural production collapsed. In the United Kingdom, the contribution to the war effort of the agricultural industry was very important. Although mechanisation had increased throughout the agricultural industry during the 1920s and 1930s with the introduction of milking machines and tractors, the shortage of labour during the war further promoted the use of machinery and the modernisation of agricultural practices. As a result of cultivation orders, the ploughing up of grassland was extensive. Between 1939 and 1945 the area of land under arable production increased by over 50% with over 2 million hectares of permanent grassland converted to arable production. In 1942, the United Kingdom Government set up the Scott Committee on Land Utilisation in Rural Areas, which was instrumental in shaping agricultural policy after the war. At this time, the agricultural depression of the 1920s and 1930s led to the belief that the main threats to the agricultural landscape was the abandonment of food production and the encroachment of urban areas. The committee envisaged the continuation of traditional mixed farming systems and thought a prosperous agricultural industry would ensure the preservation of the British countryside. This view, however, was not endorsed by the whole committee. A minority of the committee predicted a highly mechanised and specialised industry employing few people, as a prosperous agricultural industry would have to be highly efficient. The Scott report significantly underestimated the extent of the changes that would occur in the agricultural industry and the impact it would have.

After the 1939–45 war, a prosperous agricultural industry and secure food supply was seen as of strategic importance and one of the primary aims of the United Kingdom Government was to increase the productivity of the agricultural industry and balance the books (there was a huge balance of payment deficit). The 1947 Agriculture Act introduced a system of price guarantees to support agricultural production. The main objectives of the act were to 'promote and maintain ... a stable and efficient agricultural industry, capable of producing food at minimum prices, with proper remuneration and living conditions for farmers and workers'. Similar aims and approaches were adopted in Europe. The signing of the Treaty of Rome in 1958 by the six founding members of the European Economic Community supported the development of a common market in agriculture and agricultural products. The CAP aimed to increase farm productivity, secure food supplies and maintain farm incomes by offering price guarantees to all farmers.

The expansion of agricultural production

With the introduction of the 1947 Agriculture Act and guaranteed prices for several products including cereals, sugar beet, beef and milk, farmers in the United Kingdom were encouraged to increase agricultural production. Nevertheless, food shortages were still common and the rationing of food continued. Further government intervention in the industry was required to increase production. The support of farming increased through the provision of new grants and subsidies. In 1952 the Agriculture (Ploughing Grant) Act was introduced: farmers were paid £30 ha^{-1} to remove 12-year-old grass and to convert to cereal production. Subsidies were made available to farmers to increase fertilisers and lime application and grants were also introduced to remove hedgerows and improve land drainage. Food shortages finally ended in the United Kingdom in 1954 when the rationing of meat finished.

In the United States, in contrast to Europe, the problems associated with post-war reconstruction, particularly food shortages, were not an issue. The price supports introduced in 1933 to maintain farm incomes resulted in a rapid expansion of agricultural production. Surpluses in some products began in the 1950s and compulsory land diversion (set-aside) was introduced in 1956 in order to reduce supply and maintain farm incomes (Potter, 1998). Production continued to expand, however, as production intensified on the remaining land and production of crops not covered by the scheme increased. The advent of food surpluses instigated a debate between those that believed that price supports should be continued and those who wanted to end government inter-vention in the agricultural industry. A compromise was agreed in 1965 with the introduction of the 1965 Food and Agriculture Act. A system of deficiency payments set in relation to world prices was introduced (Effland, 2000).

The impetus of the modernisation of agricultural practices that started in the 1930s continued during the 1950s, as the increase in support for the agricultural industry through the introduction of price guarantees and grants combined with advances in technology to result in the expansion and intensification of

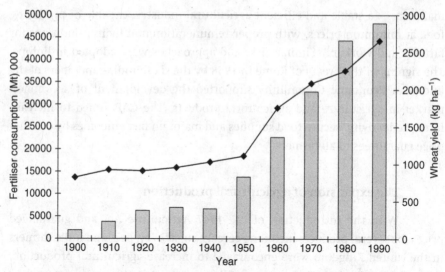

Figure 2.1 Fertiliser consumption (bars) and yield of wheat (solid line) in the United States 1900–90.

(Source: USDA, www.usda.gov/, last accessed 2006)

agricultural production. Farmers employed technologies such as artificial fertilisers, pesticides, drainage and irrigation more readily as high price supports increased production. Artificial fertilisers had been available since the late nineteenth century but had not been readily applied during the early twentieth century. The expansion in the use of machinery throughout the 1920s and 1930s and the introduction of agricultural support encouraged the increased use of fertilisers and the concomitant crop yield (Figure 2.1). The expansion of agricultural output from the 1950s was unprecedented, with crop yields more than doubling where these technological innovations had been adopted (Figures 2.1 and 2.2).

The increase in yield of cereals over this period, however, cannot just be attributed to the increase in use of fertilisers over the same period. Plant breeding and the development of new and improved varieties was also instrumental in increasing the yield of many crops. For example, old cereal varieties tended to be tall-stemmed and small-grained, which easily lodged in response to increased growth with the addition of fertilisers. The introduction of short-stemmed varieties and the relocation of stem growth to grain growth achieved increases in yield. It has been estimated that between 23% and 45% of the increases in the average yield in the United Kingdom for wheat, barley and oats are due to new improved varieties (Silvey, 1986). The expansion of agricultural production was not just limited to the arable sector. In the United Kingdom, the Committee on Grassland Utilisation formed in 1958 evaluated agricultural practices to

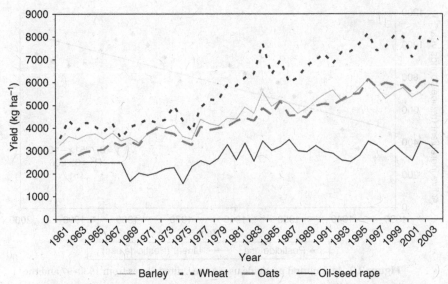

Figure 2.2 The change in crop yields from 1961–2004 in the United Kingdom.
(Source: FAO, www.fao.org/faostat/, last accessed 2006)

stimulate the production and use of grass and other green fodder crops. Silage
production was particularly encouraged. Technical advances in agronomy and
management have also contributed to increases in agricultural production by
improving the ability of farmers to maximise crop production by reducing pre-
and post-harvest losses caused by disease, pests and weeds.

The technical advances discussed above in combination with the economic
support that agriculture received in many countries increased production so
successfully that the shortage of food that was experienced in Europe in the
1950s had been converted into food surpluses by the 1980s. During this period
of technological change there was also a fundamental change in the structure of
farms, with many farmers seeing that the best way to benefit from the new
technology was to adopt increasingly more specialised management practices.
As a result, in many regions the proportion of farms that were considered mixed
farms, producing both livestock and annual crops, decreased.

Rise of environmental concerns

Historically, the environmental impacts of the agricultural industry
had not been taken into account and there were few environmental constraints
on agricultural practices. The view that a prosperous agricultural industry
would ensure the preservation of the countryside still held sway; it was generally believed that agriculture was a benevolent activity in the countryside.

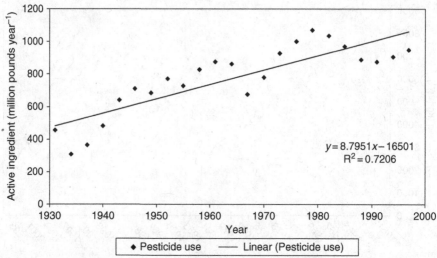

Figure 2.3 Agricultural pesticide use in the United States from 1930–97 and the linear trend line.
(Source: USDA, 1997)

Public concern over the environmental impacts of agriculture first surfaced in the United States in the 1960s, with anxiety over the impact of pesticides on both the environment and human health (Figure 2.3). During the 1970s, it was acknowledged in both the United States and Europe that the increasingly special-ised agricultural practices encouraged by the economic incentives inherent in current agricultural policy were having an adverse impact on the environ-ment. The increasing levels of pollution from intensive cereal and livestock systems and the increased use of fertilisers was no more acceptable than from any other industry. In addition, existing measures to protect habitats and wild-life were considered inadequate as the decline in habitats resulting from changes in agricultural practices continued. It was recognised that the interac-tions between agriculture and the environment were very complex but also that agriculture and the environment were interdependent. However, it was also accepted that there were many difficulties in reconciling modern agricultural practices with those of the environment.

The framework for developing policy to address the negative impact that agricultural development was having on the environment was based on an understanding of the economic incentives directing farm management deci-sions. The economic theory suggests that farmers make management decisions to maximise the profits from their land (or other productive inputs). This means that they manage the land to produce commodities that can be sold for a price in a market. However, since environmental goods and services such as biodiversity or downstream water quality are not sold in a market, there will be no economic incentive for the farmers to invest in their provision. In some cases the provision of environmental goods and services may even impose an additional cost on the farmer, for example where conserved wildlife habitat results in greater crop depredation or the setting aside of riparian buffer zones increases the fuel costs of field operations. As a result, in the absence of policy to correct the market signals, agricultural landscapes will tend to undersupply environmental goods (e.g. surface water quality) and/or oversupply environmental bads (e.g. pesticide pollution) from society's perspective. This problem has been exacerbated by pressures to intensify agricultural production in the face of increasing food demand, the introduction of production-coupled policies and changing production technology.

To correct for the individual decisions made by farmers that result in agri-environmental degradation, a range of policy approaches have been developed. These policy approaches are focused on altering the incentives such that farmers make management decisions that maintain or increase environmental goods and services. In general these incentives could take the form of compensation paid to a farmer who adopts conservation management, charging a farmer who degrades the environment an additional cost or simply regulating what farmers can and cannot do to meet some environmental objective (for a more complete discussion of these mechanisms see Chapter 4). Despite the recognition that modern agricultural practices were resulting in significant impacts on the environment and the understanding of possible policy approaches to address the problem, food production remained an important issue and the primary policy objective. It was not until the 1980s when food surpluses became prevalent and the increasing cost of agricultural policy that encouraged production, including the private and social cost of environmental degradation, became problematic that reforms to agricultural policy were initiated.

The advent of food surpluses

By 1980, Europe had achieved self-sufficiency in butter, sugar, beef and cereals. The agricultural industry had successfully fulfilled the strategic objective of a secure food supply set decades earlier. However, during the 1980s these food surpluses grew and the cost of storing and handling these food surpluses also increased enormously. To avoid farm prices collapsing surplus goods were either stored, disposed of or sold on the world markets, all with the aid of subsidies. The CAP was paying farmers to overproduce. In order to ensure a reduction in surpluses a series of adjustments to the CAP were put forward.

At this time, wholesale reform of the CAP was not seen as necessary. In 1984, in response to the increasing surpluses, a quota system was introduced across the European Union to control the output of milk. In addition modifications to the system of price support took place. Farmers were actively encouraged to transfer out of crops that were in surplus, such as wheat and oil-seed rape/canola. This was done by limiting the amount of a particular product that received full EU support. Although in some countries the amount of cereals in storage fell, further limitations on agricultural production were necessary to reduce surpluses further and to reduce the huge costs of food storage. In 1988, grant aid was made available to farmers across Europe to take arable land out of production (The Set-Aside Scheme). Under this scheme, farmers received annual payments per hectare to take at least 20% of their arable land out of production for five years, with the option of leaving the scheme after three years. Across Europe this scheme attracted little attention; in the United Kingdom very little land was taken out of production, the cereal acreage decreasing by only 3%. In 1991, this scheme was supplemented by a further initiative to make payments for one year to encourage farmers to let land lie fallow. These schemes, however, were voluntary for farmers and even though the cropped acreage had declined cereal output was still increasing. Over a decade after self-sufficiency had been achieved, the problem of food surpluses posed by the CAP still remained. In order to break the link between production and subsidies fundamental reforms of the CAP were seen as essential.

The start of agricultural policy reform

In 1992, the MacSharry reforms aimed at reducing surpluses and controlling expenditure of the CAP were adopted. These reforms changed the emphasis of agricultural support from the product (by high price guarantees) to the producer (through direct compensation payments). Measures to support prices were replaced. The main feature of the reforms was a reduction in cereal prices and the introduction of the Arable Area Payments Scheme (AAPS). Direct aid was paid on an area basis to arable producers, subject to the requirement to set aside 15% of the area on which the payment was made. Although, in theory, AAPS was a voluntary scheme, in order to maintain incomes with the accompanying reduction in cereal prices the majority of farmers had to comply. The initial AAPS was introduced as a rotational scheme with land taken out of production once in every six years. In 1993 a non-rotational option was introduced, allowing land to be taken out of production for five years. In the United States, the cost of price support to the agricultural industry was also intensely debated (Effland, 2000). The argument to reduce price supports and develop export markets so that farmers could compete on world markets gained strength and a series of government acts were passed to reform agricultural policy and shift the emphasis of price support away from production. The US Food Security Act of 1995 provided a course for US agricultural policy for the 1986–90 period. Changes from the previous Act included lowering of minimum price support levels, a decline in minimum target prices, whole dairy buy-out programmes, export enhancement initiatives and initiatives to increase farmland conservation and removal of land from production (set-aside). Although the reforms in agricultural policy were aimed at controlling government expenditures, the reforms were in part a response to the Uruguay Round of the World Trade Organization (WTO) negotiations on the General Agreement on Tariffs and Trade (GATT) which took place from 1986–94. At this time, world agricultural trade was ruled by domestic and export subsidies and import duties. The long-term aim of the negotiations was to substantially reduce subsidies and to develop a fair system of agricultural trade through reform. Agreement was reached on reducing domestic and export subsidies and import duties on agricultural products to be implemented over the six years from 1995. The agreement also committed members of the WTO to start further negotiations on continuing trade reforms in 2000.

Introduction of agri-environment schemes

Agri-environment schemes emerged as part of agricultural policy in several countries during the 1980s. In Europe, the incorporation of the environment into agricultural policy began in 1985 with the introduction of Regulation 797/85, the Agricultural Structures Regulation. The main purpose of the legislation was to restructure the agricultural industry to improve efficiency, but as part of the regulation Article 19 allowed national governments to subsidise environmental management on farms in designated Environmentally Sensitive Areas (ESAs). It was however an optional policy on behalf of member states and not all members adopted it. There was no contribution of money from CAP. Schemes were introduced in Germany, the United Kingdom, France, Greece and Denmark. The schemes implemented had wide-ranging aims, including reducing nitrogen pollution, conserving vulnerable landscapes and habitats and maintaining agricultural activities in remote areas. In 1987, a further amendment of the CAP structure, policy Regulation 1760/87, allowed a 25% contribution from the CAP.

In the United Kingdom, the government introduced the new land designation of ESAs in the 1986 Agriculture Act. The Act also imposed a legal obligation on the Minister of Agriculture to balance conservation of the countryside with a stable and efficient agriculture industry. Five areas were initially designated as ESAs in 1987 and a further five in 1988. The purpose of the designation was to 'help conserve those areas of high landscape and/or wildlife value which are vulnerable to changes in farming practice'. The detailed management requirements varied between ESAs, nevertheless, most agreements include restrictions on fertiliser use and stock densities, constraints on land improvement (such as drainage) and incentives for landscape and habitat management. However, the scheme was voluntary, so not all the farmland within each ESA was entered into the scheme. Payments were non-discretionary but there was a tiered system of payments with higher payments being made for more restrictive agreements.

At the same time as these European initiatives, policy measures to address the environmental impacts of agriculture were introduced in the United States. The 1985 Food Security Act launched a series of conservation initiatives including cross-compliance and land set-aside programmes. Cross-compliance initiatives included Sodbuster and Swampbuster, which made farmers ineligible for commodity programme benefits if sodland or swampland was tilled for crop production, and Conservation Compliance, which linked commodity programme ineligibility to inappropriate management of highly erodible farmland. Land set-aside was addressed by the Conservation Reserve Program (CRP) in which the government rented approximately 14.7 million hectares for 10 to 15 years. Conservation Reserve Program land was converted to some form of perennial vegetative cover. While the CRP was rooted in resource conservation, with highly erodible land being targeted, the initial motivation was really supply control and income support (Cain and Lovejoy, 2004). It was not recognised until some time later that the CRP could play an important role in meeting conservation and environmental objectives.

The development of environmental objectives into agricultural policy

The integration of environmental concerns into European agricultural policy, the so called 'greening' of the CAP, has been a slow and complex process (Robson, 1997). Although the primary aim of the MacSharry reforms was to reduce surpluses and control expenditure of the CAP by changing the emphasis of agricultural support, surpluses allowed the role of the farmer to be re-evaluated, and several accompanying measures were also introduced in the 1992 reforms. One of the main accompanying measures that had environment objectives was Regulation 2078/92, the Agri-Environment Regulation. The regulation involved the implementation of 'production methods compatible with the requirements of protection of the environment and the maintenance of the countryside'. The Agri-Environment Regulation consisted of seven specific objectives (European Commission, 1997):

1. reduce the polluting effects of agriculture;
2. protection and improvement of the environment, countryside and landscape, genetic resources, soil and natural resources;
3. extensification of farming and the conversion of arable land to extensive grassland;
4. the upkeep of abandoned farmland;
5. education and training;
6. land management for public access; and
7. long-term environmental set-aside.

Under this over-arching regulation each member state was required to submit a specific programme of measures to the European Commission for approval. Agri-environment measures introduced by each member state were allowed to match local conditions and could be applied on a regional or national basis. Although obligatory at the member state level, all the programmes were voluntary for farmers, with the payments based on income forgone or cost incurred. The contribution of the CAP to the programme of environmental measures increased to 50% or 75% in disadvantaged regions.

The introduction of the Agri-Environment Regulation also fulfilled several international agreements. The Convention on Biological Diversity (CBD), signed at the Earth Summit held in Rio de Janeiro in 1992, committed signatories to develop national plans for the preservation of biodiversity. This agreement had important implications within the European landscape where agricultural production occurred on approximately 40% of land surface in 2002 (European Commission, 2003a) (Table 2.1). As a result, the main approach to preserve biodiversity within Europe was to increase the area of farmed land under positive environmental management, with agri-environment measures the most widespread instrument employed to do this. In addition to national plans a European Biodiversity Strategy has been developed and action plans for areas such as agriculture established (CEC, 2001). The objective of the agriculture action plan is to maintain or improve biodiversity by promoting and supporting environmentally friendly farming practices, including measures related to genetic resources, and to prevent further loss due to agricultural activities. The United States is not a party to the CBD. However Canada ratified the CBD in 1993 and developed the Canadian Biodiversity Strategy in 1995. The Canadian Biodiversity Strategy objectives with relevance to agriculture are: (1) to conserve agricultural biodiversity which focuses on maintaining the genetic diversity of domestic and wild agricultural plants and animals; and (2) to conserve natural

Table 2.1. *The proportion of land surface under agricultural production in 2002*

Country	UAA[a] (1000 ha)	UAA (% of land surface)
Austria	3 387	40.4
Belgium	1 393	45.6
Denmark	2 690	62.4
Finland	2 216	6.6
France	29 622	53.9
Germany	16 971	47.5
Greece	3 917[b]	29.7
Ireland	4 372	62.2
Italy	15 341	50.9
Luxembourg	127	49.0
Netherlands	1 933[b]	54.4
Portugal	3 813	41.5
Spain	25 554	50.6
Sweden	3 039	6.8
United Kingdom	15 722[b]	64.7
EU – 15	130 809	40.4

Source: European Commission (2003a).
[a] Utilised Agricultural Area,
[b] 1999.

biodiversity through initiatives to fund research, extension and conservation of critical habitat.

An important consideration in the development and implementation of agri-environmental policy is whether the mechanisms used to provide environmental benefits are consistent with WTO trade rules. Under the WTO rules agriculture subsidies are categorised as either amber box, blue box or green box. Amber box measures include all domestic support measures that are considered to distort production and trade including measures to support prices or subsidies directly coupled to production (WTO, 2006). World Trade Organization members are committed to reduce amber box subsidies. Blue box measures include any support that would normally be amber box but the support also requires limits to production. There are, at present, no limits to spending on blue box subsidies. Green box subsidies must not distort trade, must be government funded and must not involve price support. These programmes tend to include direct income supports for farmers that are decoupled from production levels or prices. Green box programmes also include environmental protection and regional development programmes and are allowed without limits. Agri-environmental programmes, including environmental measures in the CAP as well as the CRP in the United States, are largely considered green box. However, in WTO negotiations some member countries have argued that due to the large payment size or the nature of some agri-environmental programmes the trade distortions are not insignificant and as such may not meet green box requirements, while other member countries believe that the current criteria are adequate or may even require greater flexibility to account for environmental concerns. It is useful to note that while green box payments constitute the main category of domestic support in many Organisation for Economic Co-operation and Development (OECD) countries (e.g. 80% of total domestic support in the United States during the 1995–8 period), expenditures on environmental programmes are only a minor component (Diakosavvas, 2003). Most of the expenditures on green box policies made by OECD countries were made for domestic food aid and general services.

The primary aim of the agri-environment measures introduced across EU member states was to support environmentally beneficial farming practices, including organic farming and the maintenance of existing low-intensity systems, although considerable differences did exist between member states in the measures introduced, representing different national priorities (European Commission, 1997). For example, in Denmark, agri-environment measures introduced included programmes to reduce nitrate pollution, and encourage organic farming and extensive grassland management. In Portugal, programmes were developed to maintain traditional extensive farming systems and reduce inputs (Working Document, 1998). In the United Kingdom, the ESA scheme was expanded to cover 43 targeted areas. In addition, a range of entirely new measures were introduced, encouraging organic farming, the removal of land from arable production to promote the development of specific habitats, extensification of moorland grazing, and were available to all farmers. Implementation of the Agri-Environment Regulation varied enormously between member states (Working Document, 1998), with large differences in payment rates and proportion of land under agreement (Table 2.2). Nevertheless, by 1998, over 28 million hectares of agricultural land were entered into some form of agri-environment measure (European Commission, 2002a).

The introduction of the Agri-Environment Regulation acknowledged the link between the intensification of agricultural production and environmental degradation. The agri-environment measures implemented have been shown to have benefited the environment by reducing the abandonment of farming practices, but have accomplished very little in terms of changing intensive agriculture to environmentally sensitive practices. Targeting of the measures has also been considered to be poor or ineffective with the payment rates too low to attract many farmers to change to environmentally sensitive practices. In

Table 2.2. *Application of Regulation 2078/92 in the European Union in 1998*

Country	% UAA[a] under agreement 1998	Average payment per hectare 1998
Austria	85.0	140
Belgium	1.9	348
Denmark	4.0	142
Finland	96.4	125
France	19.8	45
Germany	34.5	83
Greece	0.9	328
Ireland	24.7	129
Italy	14.9	266
Luxembourg	81.0	82
Netherlands	2.0	268
Portugal	22.7	105
Spain	3.6	82
Sweden	86.8	68
United Kingdom	16.0	41
EU-15	21.6	99

Source: European Commission (2002a).
[a] Utilised Agricultural Area.

addition it has been specified that agri-environment payments should only be made to practices beyond those of 'good farming practice' (European Court of Auditors, 2000). One of the major problems raised is that the fundamentals of farm support remained and as a result the introduction of environmentally sensitive agricultural practices across the whole agricultural sector was limited. Agri-environmental programmes were essentially considered a by-product of agricultural policy (Bignal and Baldock, 2002).

A multifunctional agricultural industry

The European model for developing agri-environmental policy is strongly influenced by the characterisation of agriculture as a multifunctional industry. A multifunctional agriculture is one that produces not only food and fibre commodities, but also a range of non-market goods and services. These non-market goods and services include the impacts that agriculture has on environmental quality including rural landscape amenities, biodiversity and water quality as well as socioeconomic viability of the countryside, food safety, animal welfare and cultural and historical heritage (Lehtonen *et al.*, 2005). While there is not one accepted definition of multifunctionality, a working definition provides the two fundamental parts as: (1) the existence of multiple commodity and non-commodity outputs that are jointly produced, and (2) the fact that some non-commodity outputs exhibit characteristics of externalities or public goods and public bads with poorly represented market value (OECD, 2001a). It is important to highlight the characteristic of joint production in the above definition. Joint production refers to situations where a firm produces two or more outputs that are interlinked so that an increase or decrease of the supply of one output affects the levels of the others. For example where certain aspects of biodiversity are produced jointly with sheep production (e.g. hedgerows, hill pastures) an increase in sheep production will increase the supply of that aspect of biodiversity. Alternatively, an increase in the production of an annual crop may jointly produce an increase in the amount of chemical pollution in an adjacent river. Therefore, in the presence of multifunctionality and jointness environmental benefits are increased (or decreased) with the level of commodity production. This provides an important part of the argument that has been put forward by Europe for supporting agricultural production to meet environmental objectives.

The nature of agri-environmental policy in Europe is somewhat different to what has developed in other regions. To some degree the difference can be attributed to the characterisation of agriculture, including multifunctionality, that underlies the policy framework. For example, North American policy makers have tended to not characterise agriculture as a multifunctional industry. In the absence of multifunctionality, with respect to environmental goods and services, it is not necessary to support agriculture to meet environmental objectives. It has been stated that US agri-environmental policy has traditionally treated agricultural production and the environment as substitutes, such that there is a conflict between the goals of maintaining or expanding agricultural production and preserving the environment (Baylis et al., 2003). In contrast EU policy treats agricultural production and the environment as complements such that the expansion of agricultural activity can benefit the environment, provided that it is undertaken in an appropriate manner. Supporting this characterisation is the historical trend that in North America agri-environmental policies focus on removing parcels of land from agricultural production through rental or easement agreements (Figure 2.4). For example, the CRP targets removing land from annual crop production to provide, among other environmental goods and services, wildlife habitat and biodiversity, nutrient and pesticide pollution reduction in surface water and reduction of soil erosion. Under multifunctionality these goods and services would be delivered by encouraging specific agricultural management practices. However, it should be noted that in the

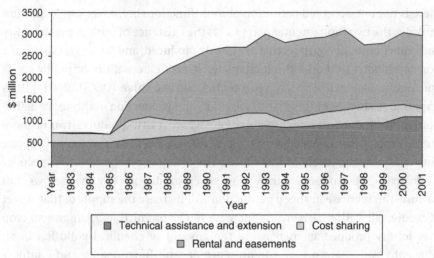

Figure 2.4 United States conservation expenditures by activity expressed in constant dollars (1996)
(Source: USDA, 2003a)

most recent US policy reforms there has been increasing emphasis on agri-environmental programmes that target 'working lands'. This may point to an acknowledgement of agriculture being a multifunctional industry in some respects and will be discussed in more detail later in this chapter.

Rural development: the European agricultural model

Further reform of the CAP was put forward in the 'Agenda 2000' proposals. The aim of Agenda 2000 was to establish a policy framework and budget for the European Union from 2000–6, in light of various challenges and pressures. Firstly, the next round of WTO negotiations due to start in 2000 was likely to increase the pressure to reduce production subsidies further. Members were committed to continuing reforms in agricultural trade through increasing trade liberalisation. Secondly, the expansion of the European Union to 25 members in 2004 and the increasing pressure this was likely to have on the budget. The proposals for the agricultural sector aimed to further strengthen the reforms began in 1992 by replacing price support measures with direct aid payments and introducing a rural policy for Europe. This reform in policy recognised that agriculture not only produces agricultural products but also plays an important role in supporting the rural economy and community and in conserving the countryside.

Reforms of the CAP under Agenda 2000 first took place in 1999. As part of these reforms the CAP was reorganised into two policy areas: market policy (known as the 'first pillar') and sustainable development of rural areas (known as the 'second pillar'). As part of these reforms, the Rural Development Regulation, Regulation 1257/1999, was introduced and the integration of environmental concerns into agricultural policy first introduced in the 1992 reforms continued (Bignal and Baldock, 2002). The Rural Development Regulation achieved further integration of environmental objectives into agricultural policy as it applied to measures under both the first and second pillars of the CAP. Under the first pillar, direct payments for agricultural products were subject to environmental requirements. Member states could comply with this aspect of the measure through three different mechanisms. Member states could either require farmers to undertake agri-environment measures or to meet specific environmental targets as a condition for payment (cross-compliance), or set compulsory environmental objectives (by introducing legislation). If farmers failed to comply with the environmental requirements then payments could be reduced or even withdrawn. Measures laid out in the Agri-Environment Regulation were incorporated into the Rural Development Regulation under the second pillar of CAP, to form a compulsory measure within that of rural development policy.

The change in policy aimed to place agricultural support in a much wider rural context. The key objective of the Rural Development Regulation was to support rural areas and not just farming. The Rural Development Regulation introduced 22 measures. Member states could choose which measures to implement according to the specific needs of their own rural areas. The measures introduced fall into seven broad categories:

1. agri-environment measures;
2. human resources: young farmers, early retirement and training;
3. less favoured areas and those subject to environmental constraints;
4. investments in farm businesses;
5. processes and marketing of agricultural products;
6. forestry;
7. promoting the development of rural areas.

Priorities in rural development policy varied considerably between member states. For many countries including the United Kingdom, Italy, Denmark, Sweden and Belgium agri-environment measures were the main priority. In other countries such as Spain, Germany, Greece and the Netherlands promoting rural development was the main priority (Table 2.3). The Netherlands chose to spend over 74% of the rural development budget on promoting the development of rural areas, including land reparcelling, water resource management and rural infrastructure. As previously mentioned agri-environment measures were

Table 2.3. *The proportion of rural development budget allocated to agri-environment measures 2000–6 and the main priorities*

Country	Agri-environment measures (%)	First priority	Second priority
Austria	64.0	Agri-environment	Less favoured areas
Belgium	41.0	Agri-environment	Investment
Denmark	38.0	Agri-environment	Forestry
Finland	40.0	Agri-environment	Less favoured areas
France	15.0	Less favoured areas	Marketing
Germany	31.0	Rural areas	Agri-environment
Greece	6.6	Rural areas	Human resources
Ireland	46.8	Agri-environment	Investment
Italy	31.0	Agri-environment	Rural areas
Luxembourg	46.0	Agri-environment	Less favoured areas
Netherlands	13.6	Rural areas	Agri-environment
Portugal	17.3	Agri-environment	Investment
Spain	9.5	Rural areas	Forestry
Sweden	50.0	Agri-environment	Less favoured areas
United Kingdom	34.0	Agri-environment	Less favoured areas

Source: European Commission (2003b).

the only compulsory component of rural development policy. Member states were required to apply these measures in their rural development programmes although they remained voluntary for farmers. This signified the importance attached to agri-environment measures in fulfilling environmental objectives, such as the Biodiversity Action Plan for agriculture. However the priorities of member states apparent in 1998 (Table 2.2) continued with the introduction of the Rural Development Regulation and the allocation of funds to agri-environment measures (Table 2.3).

As part of the Rural Development Regulation the concept of minimum environmental standards or 'good farming practice' was also introduced; each member state was required to develop a code for good farming practice applicable to the types of farming within each country, at either a regional or national level. The codes covered wide-ranging issues including soil management, water use, fertilisers, pesticides, biodiversity and landscape, pasture management and waste management. Farmers would only qualify for payments under agri-environment schemes if agricultural practices surpassed what is considered good farming practice. This ensured that the measures included in agri-environment measures delivered greater environmental benefits. Farmers were also ineligible for support under several rural development measures, including farm investment and young farmers, unless these minimum environmental conditions were met.

In many countries the agri-environment measures set up under Regulation 2078/92 continued as before under the new Rural Development Regulation (France was one of the exceptions to this with the introduction of a new scheme). As a consequence the criticisms that agri-environment measures were poorly targeted, have low payment rates and have accomplished very little in terms of changing intensive agriculture to environmentally sensitive practices largely remain unanswered. However, the adoption of the concept of minimum environmental standards and the implementation of the code of 'good farming practice' will ensure that agri-environment payments will only be received by farmers for additional environmental benefit. Further, the direct payments farmers receive for agricultural products under the first pillar are expected to achieve environmental benefits as failure to comply with the environmental requirements could result in the payments being reduced or even withdrawn.

By 2002, over 24% of the farmed area in the European Union was entered into some kind of agri-environment measure (European Commission, 2003b). Agri-environment measures involving the reduction of inputs (extensification) covered the greatest area of land (11.4 million hectares), followed by measures aimed at biodiversity and landscape (8.1 million hectares). However, this varied greatly between countries. For example in Germany, Finland and Luxembourg extensification measures are the most important, in contrast to the United Kingdom, Sweden and France, where biodiversity and landscape measures are the most important. Organic farming is the most important measure in Denmark (European Commission, 2003b).

Spending on agri-environment measures has increased greatly with 30 000 million euros, some 10.2% of the CAP budget, allocated to rural development programmes from 2000–6 (European Commission, 2002b). Given the large amounts of money that are being spent on agri-environment measures across Europe, it is very important that agri-environment measures are successful in delivering biodiversity enhancement and other environmental goods. The shift in emphasis in agricultural policy and in the manner in which farmers are supported relies heavily on the assumption that society is willing to pay for the environmental benefits produced and maintained by agricultural practices. The major reforms of the CAP in 1992 and 1999 have integrated the environmental and rural development concerns of the public into agricultural policy, as environmental priorities have to be taken into account in both the first and second pillars of the CAP. The public, however, perceives further reforms of agricultural policy necessary. A majority of Europeans want to see changes in the way the CAP supports farmers (European Commission, 2003c). For the public, the main priorities of the CAP should be to ensure that agricultural products are healthy and safe, promote the respect of the environment, protect medium- or small-sized farms and help farmers to adapt their production to consumer expectations.

Agri-environmental policy in the United States and other jurisdictions

In the United States, the first agricultural policies that specifically targeted such environmental objects as water quality and soil erosion were those that were implemented with the 1985 Food Security Act (Conservation Compliance, Sodbuster, Swampbuster and Conservation Reserve). Following these initial policies the 1990 Farm Bill implemented additional agri-environmental initiatives including: (1) Integrated Crop Management (ICM) – provided funding to farmers to cover costs associated with adopting such practices as pest and nutrient management, cover crops and improved rotations for resource conservation; (2) Water Quality Incentive Program (WQIP) – provided financial assistance to farmers for adopting management practices that conserved water quality; (3) Wetland Reserve Program (WRP) – paid farmers a rental fee for converting farmland into wetlands. In the 1996 Farm Bill the CRP was extended and the ICM and WQIP were combined and expanded into the Environmental Quality Incentives Program (EQIP), which was set up to provide five- to ten-year cost-share or incentive payment contracts to crop and livestock producers for specified conservation management programmes. The EQIP began with $200 million in annual funding of which half was earmarked for livestock producers. The 1996 Bill also established a new Wildlife Habitat Incentives Program (WHIP) to induce wildlife habitat reclamation from production land. At about the same time Conservation Compliance was weakened under farm lobby pressure with farmers feeling it too intrusive (Cain and Lovejoy, 2004). The targeting of agri-environmental policy changed with the 1996 Farm Bill from an approach that attempted to have all areas participating to an approach that 'maximized environmental benefits per dollar expended'. Programmes were targeted to conservation priority areas, which functioned to funnel conservation dollars away from the general farming public into areas that were classified as environmentally critical. However, the 1996 Bill was only marginally successful in attaining benefit-cost targeting of conservation dollars (Cain and Lovejoy, 2004).

The 2002 US Farm Bill continued the focus on conservation with funding for environmental programmes being increased 80% over the 1996 Farm Bill. For example, EQIP funding increased from $200 million to $1.3 billion over several years and a new Conservation Security Program (CSP) was established. The CSP paid producers to adopt or maintain practices that address soil, water and wildlife concerns. The CSP was set up as a three-tier system with higher tiers requiring greater conservation effort but offering greater payment levels. The EQIP and CSP are considered 'working lands' programmes, which provide funding to farmers for conservation activities on lands that are used for crop production and grazing. The 2002 Bill has an increased emphasis on conservation on working lands with expenditures increasing from 15% of federal agricultural conservation programmes in 1985 to 50% of the much larger total conservation spending by 2007. Therefore, the emphasis is changing, to some extent, from land retirement to working land conservation.

> Land retirement programs have succeeded in improving environmental quality by removing the most fragile land from production, but these benefits come at a high cost to taxpayers. Moreover, now that the most fragile land has already been retired through programs like the CRP, the remaining land eligible for retirement may have higher production potential than the retired land and, therefore, may be more costly to retire. Keeping the land in production and funding conservation practices on that land may be a more cost-effective option. (USDA, 2003b)

However, with funding being recently redirected to disaster assistance for flood-impacted regions only 14% of the eligible acres had been funded up to 2004. The 2002 Farm Bill did expand land retirement programmes with increased emphasis on wetlands. The CRP area increased from 14.7 to 15.9 million hectares with a further 0.5 million hectares added to the WRP and the creation of the Grassland Reserve Program to assist landowners in restoring and conserving grasslands. Further the WHIP received a tenfold funding increase over the 1996 Bill and the Farmland Protection Program, a programme providing funds to local groups to purchase easements to protect against development of productive farmland, received increased funding. However, it should be noted that with increased expenditures on defence and homeland security expenditures on these programmes have been smaller than proposed in the Bill. The 2002 Farm Bill included a fundamental change in targeting conservation expenditures – the objective was no longer to simply maximise the area in conservation but to target based on benefit-cost criteria (maximise the environmental benefits for the expenditures) (Cain and Lovejoy, 2004).

In other developed countries, such as Australia and Canada, agricultural policy in general, and agri-environmental policy more specifically, is comprised of measures that are less focused on providing financial incentives for specific conservation management practices. For example, Australia, Canada and New Zealand were singled out by the OECD (2003a) as countries that have placed emphasis on the use of community-based approaches (employing collective action) to address environmental issues. For example, in Australia, Landcare, a national voluntary community programme initiated with government financial support in the range of AUS $40 million per year, was aimed at improving natural resource management practices. Landcare currently involves 40% of farmers who manage 60% of the land and 70% of the diverted water (Australian Government website). Further, the Australian Government Envirofund provides grants to communities to undertake local projects aimed at conserving biodiversity and promoting sustainable resource use. Australian agri-environmental policy has a strong focus on water quality and quantity issues and soil salinity problems and the control of weeds and invasive species.

In 2003 the Agriculture Policy Framework (APF) was developed by the Canadian Government. The APF identifies the environment as one of the five pillars of agricultural policy and acknowledges that in order to 'step up the pace' of addressing environmental challenges on the farm producers will require financial assistance. As a result CDN $100 million have been allocated to providing assistance to farmers for the development of Environmental Farm Plans and environmental scans to identify high-risk areas to assist the implementation of the Farm Plans. The APF proposes to use these Farm Plans to target incentives to encourage conservation management that addresses identified environmental risks. Other initiatives include incentives aimed at conversion of land from annual cultivation to grass, pesticide-risk reduction plans, sustainable water use and supply expansion plans. The government is also focused on developing a suite of agri-environmental indicators to direct future policy and is scanning agricultural policy to determine environmental implications. With Canada ratifying the Kyoto Protocol in early 2005 an important component of recent Canadian agri-environmental policy development has been climate change mitigation, with programme objectives of reducing greenhouse-gas emissions in the farm management areas of soil, nutrients and livestock management. In contrast, Australia has not ratified the Kyoto protocol and climate change issues in agriculture are more focused on climate change adaptation with mitigation initiatives being acceptable only when no economic disadvantage is imposed on the agriculture industry.

In 1986, New Zealand rapidly modified their agricultural policy framework by removing all policies that alter production or trade patterns, with levels of producer support the lowest of all OECD countries. For example, producer support in 2002–4 was estimated to be 2% in New Zealand, compared with approximately 16% in the United States, 22% in Canada and 35% in the

European Union (OECD website). Agriculture programmes tend to provide payments only for pest control, with a particular focus on biosecurity preventing the importation of exotic pests and diseases, or relief against climate disasters. As such, agri-environmental policy measures are very limited within New Zealand with some recent developments in the area of climate change research and water quality and water use policy (OECD website).

Summary

Throughout the last century agriculture has been an important industry to most developed countries, and, as a result, agriculture has been an important policy priority. The objectives of agricultural policy have changed over the years from production enhancement, income support and stabilisation and, lately, through agri-environmental policy programmes, the provision of environmental benefits and/or decreasing environmental costs. The nature of the agri-environmental programmes has been influenced by the types of environmental benefits (or environmental costs) that are deemed a priority by society and the government, the characterisation of the agricultural industry, including the presence of multifunctionality and jointness, the rules associated with international trade and environmental agreements. Finally, it is apparent that the level of commitment to agri-environmental programmes given by governments is dependent on the budget priorities and the relative importance of environmental issues compared with other rural development and industry initiatives.

3

Environmental impacts of agriculture

Introduction

It is widely recognised that agriculture plays a pivotal role in managing and maintaining landscapes around the world. However, it is also commonly accepted that both the expansion and the increasing intensity of modern agricultural practices have had a huge impact on the natural environment. In many parts of the world traditional low-input low-output farming systems remain. In order to understand the effects of intensive farming practices on the environment, we need to comprehend the changes and advances that have taken place in agricultural practices, which have transformed many of the traditional farming systems to a system of intensive monocultures. While the causal agents of environmental degradation (such as the increased use of fertilisers and pesticides) are very well known and documented, it is difficult to separate the influences of agricultural policy and new technology as the underlying drivers.

The traditional integrated low-input low-output system

Before the development and widespread use of artificial fertilisers and pesticides and large-scale irrigation and drainage schemes, crop and livestock production was dependent on the productivity provided by natural environmental conditions. Agricultural production was limited by the availability of soil water; the natural fertility of the soil; and pests, weeds and diseases. The level of inputs available to farmers was low and the level of outputs was correspondingly low. Farmers maintained the fertility of the soil and controlled pests and diseases by using a variety of different farming systems. Shifting and fallow cultivation was often used to maintain soil fertility and to control pests

42

and diseases. The interval between crops was based on the time necessary to build up the natural supply of nutrients. Permanent agriculture with continuous cultivation of the same piece of land was only possible if soil fertility could be maintained and the build-up of pests, weeds and diseases controlled. In many parts of the world, farming systems based on a rotation of different crops evolved to meet these challenges. Farmers rotated different crops on the same piece of land often using a fallow period within the rotation to control pests, weeds and diseases. The crops used by farmers in the rotations differed geographically in accordance with climate and soil type although most rotations include a cereal crop, such as wheat, maize, barley and sorghum; a legume such as peas and beans; along with a root crop such as potatoes, sweet potatoes and cassava. Cereals tended to be the most important crop within the rotation, providing a highly nutritious food source. In addition to providing a good source of protein, leguminous species were also an essential part of the rotation for the reason that they are able to fix soil nitrogen allowing a build-up of soil fertility. Animal manure also provided a valuable source of nutrients and in many farming systems production could only be maintained with the input of nutrients from animal manures and the integration of livestock and crop production. In the past, agricultural production was chosen to suit local environmental conditions, and in many parts of the world crop and livestock production is still dependent on the natural environmental conditions, and low-intensity integrated farming systems remain. Nevertheless, there has been widespread abandonment of these traditional farming systems, especially in North America and Western Europe, as a consequence of the intensification of inputs. Intensification has altered the environment and facilitated the adoption of new production systems. As a result traditional farming systems have been replaced with increasingly specialised types of modern intensive farming systems.

Technology: the driver of agricultural modernisation

New technology has been very influential in the development of highly specialised intensive farming systems with the creation of artificial fertilisers crucial to this change. By the late nineteenth century the agricultural industry was changing rapidly. The discovery of the Haber–Bosch process meant that artificial fertilisers were available to farmers and subsequently the level of inputs (the supply of nutrients) could increase. Basic slag, which contained lime and phosphate, and ammonium sulphate, two industrial by-products, were commonly used. For the first time, inputs were available from outside the farming system. As a result of this development, farmers no longer required the input of animal manure from livestock to maintain agricultural productivity, and an integrated farming system including both crop and livestock production was no longer necessary. This led to a simplification of cropping systems and the polarisation of farming systems as livestock and crop production became increasingly specialised (Stoate, 1995).

The early part of the twentieth century saw huge technological advances, with the mechanisation of agriculture and further developments in fertilisers, plant nutrition and crop varieties. These fertilisers helped agriculture expand onto previously unproductive land, increasing the area of land dedicated to annual crop production. Mechanisation allowed the simplification and concentration of farming operations into a shortened period of time. Mechanisation also enabled the expansion of agriculture as pasture-land was converted to arable production as the use of draft animals in agriculture declined and feed and fodder crops could be replaced by more profitable food crops. Although there had been huge advances in technology, uptake by farmers was by and large gradual due to the agricultural depression of the 1920s and 1930s. Animal manures remained the chief source of nutrient input until mass mechanisation took place in the 1940s.

The impact of agricultural policy on agricultural practices

As described in Chapter 2, a prosperous and productive agricultural industry was believed to be vital following the 1939–45 war. Many governments introduced systems of price guarantees and grants to support and modernise agricultural production and secure food supplies. Contrary to the popular opinion of the 1940s, the increased support for the agricultural industry, in combination with further technological advances, encouraged both the intensification and specialisation of agricultural production, and consequently has resulted in huge changes in the countryside and the natural environment (Bowers, 1985; Sheail, 1995). The expansion of agricultural output from the 1950s was unprecedented as the system of price guarantees and grants supported increased agricultural production and the increased use by farmers of technologies such as artificial fertilisers, pesticides, drainage and irrigation. These inputs continued to increase rapidly until the advent of food surpluses and the subsequent change in agricultural policy that took place during the 1980s (Figures 3.1 and 3.2).

The appearance of more intensive and specialised forms of agricultural production resulting from input and output price incentives inherent in agricultural policy has not only resulted in increased inputs but has also led to huge changes in land use and cropping patterns. For example, in England, the balance

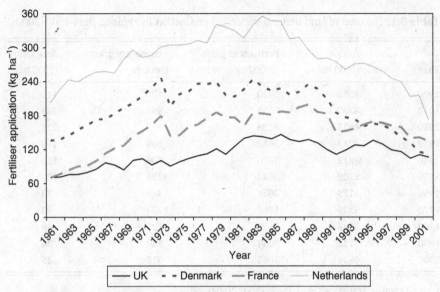

Figure 3.1 Fertiliser consumption from 1961–2004 in four European countries.
(Source: FAO, www.fao.org/faostat/, last accessed 2006)

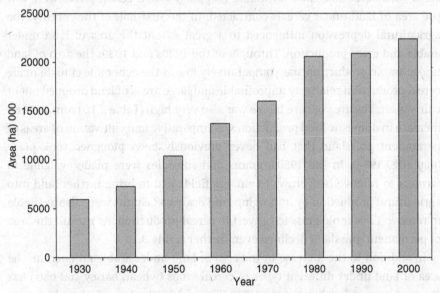

Figure 3.2 The area of irrigated land in the United States from 1930–2000.
(Source: USDA, www.usda.gov/, last accessed 2006)

between the area of land under arable, permanent and temporary grass has changed considerably throughout the last century (Table 3.1). The price of cereals influenced the area of land under arable and grass production. When the cereal prices were high the area of arable land was able to expand to those

Table 3.1. *The area of land under agricultural production in England from 1900–2000*

Year	Arable (1000 ha)	Permanent grass (1000 ha)	Temporary grass (1000 ha)	Bare fallow (1000 ha)
1900	4582	5420	1120	119
1910	4287	5635	955	139
1920	4525	5126	875	225
1930	3713	5453	860	118
1940	3812	5076	706	120
1950	5209	3648	1236	102
1960	5179	3658	1543	73
1970	5329	3261	1234	92
1980	5242	3155	1029	49
1990	5123	3054	830	34
2000	4634	2864	629	25

Source: Farming Statistics Team, Defra (2002–2004)

areas considered marginal for arable cropping. When cereal prices were low the area of land under cereals contracted. In the first half of the century, the agricultural depression influenced to a great extent the area of land under arable and grass production. Throughout the 1920s and 1930s the area of land under arable production was comparatively low as the economic climate made cereal production relatively unprofitable and large areas of land dropped out of cultivation. The area of bare fallow was also very high (Table 3.1). From 1940 an increase in domestic food production was imperative and cultivation of areas of permanent grassland that had never previously been ploughed took place (Ratcliffe, 1984). In the 1950s, grants and subsidies were made available to farmers to remove hedgerows to enlarge fields and to bring further land into agricultural production by improving land drainage. Grants were also available to remove 12-year-old grass to convert to cereal production. As a result the area of permanent grassland declined even further (Table 3.1).

In addition to the increase in the area of land under arable production, the area of land under different types of cereal crops (wheat, barley and oats) has changed tremendously since the 1960s (Figure 3.3). The area of wheat has more than doubled, following firstly a decline in the area of oats and then in barley from the 1980s onwards (Figure 3.3) and better wheat varieties becoming available. Price incentives have been an influential factor in this change in the types of cereal crops grown. In addition, price incentives have been important in the introduction of new crops. On joining the European Economic Community in 1973, oil-seed rape/canola qualified for agricultural support and as a result

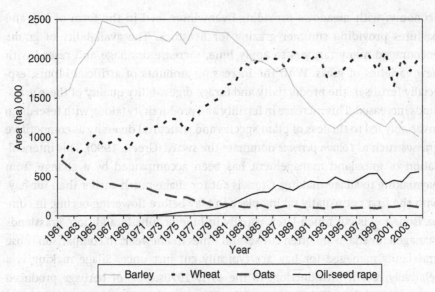

Figure 3.3 The change in cropped area from 1961–2004 in the United Kingdom.
(Source: FAO, www.fao.org/faostat/, last accessed 2006)

began to appear in the British countryside, increasing enormously in area in the subsequent 30 years (Figure 3.3).

The change in type and balance of crops grown masks another change in cropping patterns that has occurred since the 1970s: the change from spring-sown to winter-sown barley. Although the vast majority of wheat sown in the United Kingdom has been winter-sown (in 1920, only 4% of wheat was spring-sown, Defra, 2003), the preference of farmers to sow spring barley remained until the 1970s. Although winter-sown varieties were available to the farmer, there were doubts about their suitability for the British climate. In 1980, 40% of the barley sown was winter barley (Defra, 2002a) by 1990 this proportion had increased to 72% (Defra, 2002b).

The huge decline in area of barley sown from the peak in the late 1960s (Figure 3.3) together with the switch from spring-sown to winter-sown barley has resulted in a vast reduction in area of spring-sown barley grown. Subsequently the availability of stubbles to wildlife throughout the autumn and winter months has also decreased. In addition to the changes in cropping patterns, the area of bare fallow has decreased enormously (Table 3.1), as the increased use of pesticides and herbicides has reduced the need for crop rotations to control pest species and diseases. This together with the use of fertilisers has allowed farmers to continuously sow the same crop on the same piece of land.

The management of grasslands has also changed in the last 60 years (Vickery *et al.*, 2001). Meadows and pastures were a very important part of the farm economy, with meadows providing over-winter feed in the form of hay and pastures providing summer grazing for livestock. The availability of grants encouraged many farmers to apply lime, increase drainage and reseed with new varieties of grass. With the increasing amounts of artificial inputs, especially fertiliser, the productivity and forage digestibility quality of these grasslands increased. This increase in fertility and productivity (along with reseeding) invariably led to the loss of plant species and structural diversity as competitive grasses such as *Lolium perenne* dominate the sward (Green, 1990). This intensification of grassland management has been accompanied by a change from haymaking to silage making. Grass is cut for silage much earlier than for hay, with the first cut usually taking place in May before flowering occurs. In comparison, hay is cut much later, usually in June or July. In addition, grasslands managed for silage are often cut several times in one year; in comparison those grasslands managed for hay are typically cut just once. Silage making is a relatively recent introduction. In the early 1970s, 85% of herbage produced within the United Kingdom was hay. However, by the mid 1990s this had declined to 30% (Vickery *et al.*, 2001). Silage making has become more popular because the timing of cutting operations is less critical than for hay, as silage can be conserved at much greater moisture content, and because it was considered to produce a higher quality storable food.

The transformation from low-intensity to highly intensive and specialised farming systems as a result of agricultural policy in addition to advances in agricultural technology and management has undoubtedly increased agricultural production, but at what cost? The herbicides, pesticides and fertilisers increasingly used in agricultural production are known to have resulted in huge impacts on the natural environment, and the changes in agricultural management, such as cropping patterns and the timing of management operations, have been critical in the decline of many mammal, bird, invertebrate and plant species.

Environmental pollution from agriculture

Agriculture is known to be a major polluter and as a result has had a huge impact on the natural environment, on the quality, and in some cases quantity, of air, water and soil. One of the important environmental impacts agriculture has on the natural environment is the major contribution it makes to global greenhouse-gas emissions and consequently global warming. In terms of their contribution to relative warming, the three most important greenhouse gases are carbon dioxide (CO_2), methane (CH_4) and nitrous oxide (N_2O) respectively and agriculture is responsible for significant emissions of all three gases.

Table 3.2. *Atmospheric concentration of the three main greenhouse gases in 2000*

Gas	Atmospheric concentration	% Increase since 1750
Carbon dioxide	368 ppm	$31 \pm 4\%$
Methane	1750 ppb	$151 \pm 25\%$
Nitrous oxide	316 ppb	$17 \pm 5\%$

Source: IPCC (2001).

Table 3.3. *Total emissions of methane and nitrous oxide in 2004 (Tg CO_2 equivalent) and the proportion resulting from agricultural activities*

Country	CH_4	Agriculture	N_2O	Agriculture
Australia	123.7	58.1%	25.8	82.4%
Canada	110.0	24.7%	44.0	63.6%
Denmark	0.27	64.9%	0.02	82.5%
Germany	2.44	44.9%	0.21	63.6%
Netherlands	0.83	50.1%	0.06	53.7%
Spain	1.74	62.5%	0.10	76.0%
United Kingdom	2.46	36.0%	0.13	65.0%
United States	556.73	28.8%	386.71	72.3%

Source: UNFCCC (2006).

Atmospheric concentrations of all three gases have increased substantially since 1750, with the greatest increase observed in the concentration of methane (Table 3.2).

Emissions of carbon dioxide from agriculture are primarily the result of land-use change rather then the direct result of agricultural activities. Deforestation and the loss of soil carbon contribute a large proportion of the increase in carbon dioxide to the atmosphere. However, a substantial proportion of the total emissions of methane and nitrous oxide are the result of agricultural activity (Table 3.3). Wetland or paddy rice cultivation and ruminant livestock are responsible for significant emissions of methane, while nitrous oxide emissions are largely due to the use of inorganic fertilisers released from the soil through the process of denitrification (see nitrogen cycle in Chapter 7).

The increasing use of nitrogen fertilisers over the past 60 years has not only contributed to global warming and the increase in nitrous oxide recorded in the atmosphere but has also resulted in an excessive amount of nitrogen applied over large areas of agricultural land. In many parts of Europe and North America

Table 3.4. *Critical load of nitrogen – deposition below this level will result in no harmful long-term effects*

Habitat type	Critical load (kg N ha^{-1} yr^{-1})
Moorland	15 .
Heathland	17
Calcareous grassland	50
Neutral/acid grassland	25
Deciduous woodland	17

Source: NEGTAP (2001).

there is a surplus of nitrogen, measured by the imbalance between the amount of inputs and outputs. This surplus is usually lost to the environment. In Europe, the surplus of nitrogen is greatest in the livestock breeding areas, particularly regions of Belgium, Denmark and the Netherlands, where the surplus is often greater than 170 kg N ha^{-1} yr^{-1} (Schrøder, 1985; CEC, 2002). In America, often only half the applied nitrogen is removed in agricultural production leaving a surplus of over 100 kg N ha^{-1} yr^{-1} (Power and Schepers, 1989). For that reason, atmospheric deposition of nitrogen is a major problem to many natural and semi-natural ecosystems (Bobbink *et al.*, 1998). For many important habitats, the levels of nitrogen deposition they are receiving are greater than the calculated critical loads (Table 3.4).

One of the main sources responsible for the increased atmospheric deposition of nitrogen is ammonia, and agriculture, particularly livestock production, is known to be the main source of ammonia emissions to the atmosphere. In the United Kingdom, it has been estimated that 85% of the total ammonia emissions are from agricultural sources (Defra, 2002c). Of this, livestock production is the main polluter, responsible for 90% of the emissions of ammonia from agricultural sources. The increased application of nitrogen fertilisers to grassland and subsequently into animal feed translates into increased amounts of nitrogen in the dung and urine, with manure spreading and livestock housing being the main sources of ammonia pollution. The problem of ammonia emissions and their effect on the natural environment was addressed in the UNECE (United Nations Economic Commission for Europe) Gothenburg Protocol signed in 1999. Under this agreement countries agreed to decrease their ammonia emissions, with those countries with the greatest environmental impact making the biggest cuts. Consequently, countries such as Belgium, Denmark and the Netherlands whose agricultural industry is largely based on livestock production have to reduce their emissions

Table 3.5. *Ammonia emissions in 1990 (thousand tonnes per year) and the 2010 target amount agreed to under the Gothenburg Protocol in 1999*

Country	1990 Emissions	2010 Target	Percentage reduction
Belgium	107	74	−31
Denmark	122	69	−43
France	814	780	−4
Germany	764	550	−28
Greece	80	73	−9
Italy	466	419	−10
Netherlands	226	128	−43
Poland	508	468	−8
Sweden	61	57	−7
United Kingdom	333	297	−11

Source: UNECE, www.unece.org/env, last accessed 2006.

greatly compared with other countries such as France, Greece and Poland (Table 3.5).

Although ammonia emissions have been decreasing since 1990 in the United Kingdom (Figure 3.4), thought largely to be the result of decreases in livestock numbers and fertiliser use (Defra, 2002c), they have yet to reach the target of 297 thousand tonnes outlined in the UNECE Gothenburg Protocol (Table 3.5). In addition a significant increase was recorded between 2003 and 2004 (Figure 3.4). Further reductions in the numbers of livestock, thought to be a likely consequence of Common Agricultural Policy of the European Union (CAP) reform, and fertiliser use are likely to reduce emissions. Nevertheless to achieve the required reductions to meet the 2010 target set by the protocol it is likely that farmers will have to manage manure and slurry more effectively to minimise emissions. As a result, restrictions on the methods of application such as broadcasting manure and slurry may as a result be introduced. Emissions of ammonia are not the only form of nitrogen pollution resulting from agricultural activities to enter the atmosphere. Nitrogen in the form of nitrogen oxides (NO_x) largely emitted by machinery (tractors, combines) is also a source of atmospheric pollution although the contribution of agriculture is small compared with industry and road transport.

Atmospheric deposition of nitrogen is a major problem as it increases the availability of nitrogen and can also lead to acidification (Bobbink *et al.*, 1998). Atmospheric deposition rates vary considerably between and within countries. Current levels of atmospheric deposition in the United Kingdom average at about $17\,kg\,N\,ha^{-1}\,yr^{-1}$ although in some areas it is above $50\,kg\,N\,ha^{-1}\,yr^{-1}$

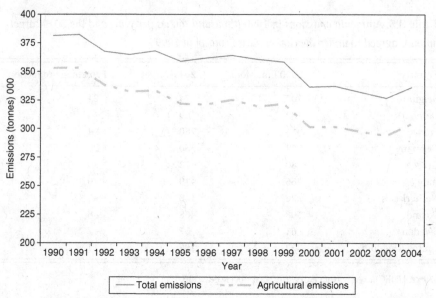

Figure 3.4 Emissions of ammonia (excluding natural sources) in the United Kingdom. (Source: National Environment Technical Centre, Defra, 2006)

(NEGTAP, 2001). The effects of increased nutrient availability on species diversity and plant community composition have been well documented in heathland and grassland ecosystems across Europe. For example, the decline of heather (*Calluna vulgaris*) can be attributed in part to the increase in nitrogen availability (Heil and Diemont, 1983; Aerts *et al.*, 1990) although other factors including overgrazing by livestock are also considered important (Bardgett *et al.*, 1995; Alonso and Hartley, 1998). For a detailed review of the wide-ranging impacts of atmospheric nitrogen deposition on different biological systems refer to Bobbink *et al.* (1998).

Nitrogen is also a major source of pollution in water, as in the form of nitrate it is readily lost from the soil through leaching. Concerns over water quality and in particular the increasing amount of nitrate in water as a result of agricultural practices first arose due to human health issues, as a high level of nitrate in water is hazardous to humans. In 1980, the Drinking Water Directive 80/778/ EEC was introduced across the European Union, implementing a legal limit on the amount of nitrate allowed in drinking water. The permissible amount of nitrate in water was limited to 50 mg NO_3 l^{-1}. Although nutrients (nitrogen and phosphorus) occur naturally in water, it was recognised that increased levels of nutrients in watercourses as a result of fertiliser use (both organic and inorganic) was a major problem across Europe. For example, between 50% and 80% of the nitrates entering Europe's waters are from agricultural sources

(CEC, 2002). In order to reduce pollution from agriculture the Nitrates Directive 91/676/EEC was introduced in 1991. The aim of the directive was to protect waters against pollution caused by nitrates from agricultural sources. This included the protection of surface and ground water in addition to lakes, estuaries, coastal and marine waters. Member states had to implement a programme of measures to limit the application of nitrogen to areas considered vulnerable to nitrate pollution. Areas that had or could have a level of nitrate greater than 50 mg l^{-1} were designated Nitrate Vulnerable Zones (NVZs). Within NVZs farmers are restricted in the amount of fertilisers they can apply. Inorganic nitrogen fertiliser application is limited to the requirements of the crop and organic manure applications to 210 kg ha^{-1} of total nitrogen for the first 4 years and then decreased further to 170 kg ha^{-1} yr^{-1}. To reduce leaching, farmers are not allowed to apply organic manures during the autumn on sandy or shallow soils. All member states were to implement the directive by 1993. However, many member states were very slow to respond to the directive. By 1997, countries including Belgium, Greece, Portugal and Spain had yet to designate vulnerable zones. In 2001, all member states apart from Ireland had designated vulnerable zones in accordance with the directive, with six member states choosing to implement the directive across their whole territory. However in many cases the area designated was much lower than the area the European Commission had identified (Table 3.6). In 2000, the European Court of Justice ruled that the United Kingdom had failed to protect all surface and ground water from nitrate pollution, even though vulnerable zones had been designated in 1996. In response to this ruling the government has increased the area of agricultural land within NVZs.

Table 3.6. *The area of land designated as NVZs in 2001 (km^2) and the additional area the European Commission considered should have been identified as vulnerable zones*

Country	Area designated	% Land cover	% Additional land
Belgium	2 700	9	51
France	240 900	48	7
Greece	13 900	11	11
Ireland	0	0	9
Italy	5 800	2	29
Portugal	900	1	13
Spain	32 000	6	14
Sweden	41 000	9	10
United Kingdom	7 800	3	8

Source: CEC (2002). Austria, Denmark, Finland, Germany, Luxembourg and the Netherlands designated 100% of land cover as NVZs.

In North America, contamination of ground water by nitrates is a particular problem, as a large proportion of the population depend on ground water for their drinking supply. Nitrate pollution has been identified as one of the most serious water quality problems in North America (OECD, 2003b). Once more, controlling the amount of nitrogen inputs made by the farmer (fertilisers and manures) appears key to reducing the problem of nitrate pollution (Power and Schepers, 1989). A significant impact of excess nitrogen is the large oxygen-depleted or hypoxic zones that develop in aquatic areas that receive nitrogen run-off from agricultural land. One of the most well known of these is located in the Gulf of Mexico, a 13 to 20 000 km^2 hypoxic zone caused by excess nitrogen from agricultural land in the central United States.

Phosphorus can also be easily lost from farming systems, mostly through the loss of soil through erosion and the leaching of phosphorus. It has been estimated that 0.9 kg ha^{-1} of phosphorus is lost each year, 0.4 kg per hectare through leaching and 0.5 kg ha^{-1} as a result of soil erosion by water (Newman, 1997). In the United States it has been estimated that only 30% of the fertiliser and feed phosphorus input to farming systems is taken up and removed as output in crop and animal products, resulting in an annual phosphorus surplus of 33.6 kg ha^{-1} (Sharpley et al., 2003). In other words, the application of phosphorus is in excess of the ability of crops to use it, resulting in soil phosphorus accumulations and an increased risk of phosphorus loss in run-off. The loss of phosphorus from farming systems is seen as a particular problem as most freshwater bodies are phosphorus-limited. Increasing the amount of phosphorus can lead to many problems associated with eutrophication. Agriculture is responsible for about 50% of the phosphorus found in water in the United Kingdom (Defra, 2004). Reducing the losses of nitrogen and phosphorus to water from agricultural activities is vital to improving water quality and the aquatic environment. Agriculture is also a major source of organic pollution. Contamination of water bodies by disease-causing pathogens including *Salmonella*, *Escherichia coli* and *Cryptosporidium* from animal manures is of great concern to human health. One of the worst outbreaks of *Cryptosporidium* occurred in 1993, when over 400 000 people became sick and more than 100 people died in the United States after drinking contaminated water.

Although European legislation to control water quality was first put in place in 1980, there has been little overall change in the observed concentrations of nitrates found in many water bodies in the subsequent 25 years (Figure 3.5). Pollution of water bodies from agricultural sources is still seen as a major problem, especially by nitrogen and phosphorus as a result of the use of fertilisers and animal manures (Defra, 2002d). Within Europe, the Water Framework Directive (2000/60/EC) requires all surface water (rivers, lakes, estuaries and

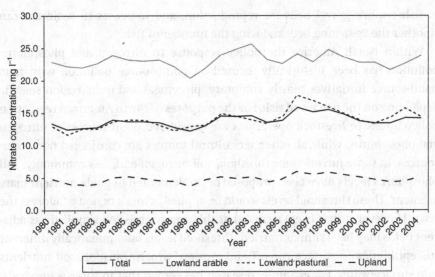

Figure 3.5 Average annual concentration of nitrates within 160 rivers in Great Britain.

(Source: Harmonised Monitoring Scheme, Defra, 2006)

coastal waters) and ground water bodies to achieve 'good ecological status' by 2015. For the first time, water quality targets will be based on ecological rather than chemical conditions and will be specific to the type of water body.

Although in recent years there has been an observed downward trend in the use of fertilisers (Figure 3.1), in order to achieve the aim of 'good ecological status' further reductions in water pollution from agricultural sources will be necessary. By 2012, all member states are required to put into practice a programme of measures to meet the requirements of the directive. This could have a huge impact on agricultural land management and practices, as failure to meet the requirements of the Water Framework Directive will result in a large fine each time the directive is contravened. It is likely that farmers, across Europe, will have to limit the amount of fertilisers and animal manures applied and implement measures to control run-off and soil erosion. In areas where the risk of pollution is very high the removal of land from agricultural production could be the only solution. As previously described, the process of soil erosion is a major factor contributing to the increase in nutrients added to water, especially phosphorus, from agricultural sources. However, the loss of soil from agricultural land to water bodies can also have a major impact on the aquatic environment resulting from the increased amount of sediment. An increase in the amount of sediment increases the turbidity of water, which has a huge impact on submerged plants (reducing their ability to photosynthesise) and subsequently the species that depend on them. In addition, many species of fish require gravel beds for reproduction, and increases in sediment can smother the spawning beds reducing the number of fish.

Within North America the policy response to nitrogen and phosphorus pollution has been historically focused on point-source pollution with non-point-source initiatives mostly voluntary, piecemeal and quite region specific depending on the perceived risk. For the purposes of North American regulation policy intensive livestock operations are considered point-sources of nitrogen and phosphorus, while all other agricultural sources are considered non-point-sources. In some jurisdictions threshold soil nitrogen and, less commonly, soil phosphorus levels have been proposed or implemented to guide nutrient management. These threshold levels would be applied across a region to address the risks. However, it has been argued that this approach is too simplistic as adjacent fields may have similar nutrient threshold levels but dramatically different susceptibilities to surface run-off and erosion, which transport soil nutrients into surface water. For example, research has shown that in some watersheds, 90% of the available phosphorus in surface water comes from only 10% of the land area during a few relatively large storms (Pionke et al., 1997). As a result, it has been proposed that nutrient run-off policy should target areas where there is both high soil nitrogen and phosphorus stocks, as well as high surface run-off and erosion potential (Sharpley et al., 2003). However, a comprehensive approach to nutrient run-off and pollution of surface water has not been implemented in the United States or Canada.

Prior to the development of synthetic pesticides farmers used crop rotations and a range of naturally occurring chemicals to control insect pests, weeds and diseases and to protect their crops. Since their introduction in the 1940s the use and number of synthetic pesticides by the agricultural industry has increased considerably (Figure 3.6) as farmers became progressively more reliant on pesticides to protect their crops within the increasingly specialised farming systems. Many pesticides are inherently dangerous and although their use in agriculture appears to have peaked (Figure 3.6) they have had a considerable impact on the natural environment. However, figures of this type are difficult to interpret because over time the pesticides produced have become more targeted and more biologically active, so that a lower volume of active ingredient is required to produce the same level of pest control. In addition to limiting the amount of nitrate in drinking water the Drinking Water Directive 80/778/EEC, introduced across the European Union in 1980, implemented a legal limit on the amount of pesticides allowed in drinking water. The permissible amount of a single pesticide was limited to $0.1 \, \mu g \, l^{-1}$. The costs of removing pesticides from drinking water to make it safe for human consumption are substantial (Pretty et al., 2000).

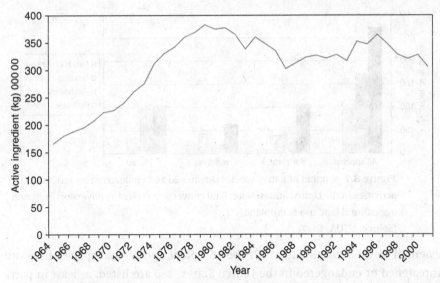

Figure 3.6 Conventional pesticide use in agriculture in the United States 1970–2001. (Source: US Environment Protection Agency, www.epa.gov/pesticides/, last accessed 2006)

The increasing use of pesticides within agriculture has also been at considerable expense to wildlife with direct and indirect effects on many species of mammals, birds, invertebrates and plants (Bunyan and Stanley, 1983). Pesticides such as DDT and Dieldrin were shown to increase in concentration throughout the food chain. Many top-predator species including sparrowhawks (*Accipiter nisus*) and kestrels (*Falco tinnunculus*) declined in number as the pesticides accumulated in their eggs, affecting their ability to reproduce successfully. There has also been concern about the drift of pesticides onto uncropped areas such as field boundaries and adjacent semi-natural habitats (Marrs *et al.*, 1989) as they are important features within the agricultural landscape and play an important role in providing habitat for wildlife.

The impact of agriculture on wildlife

Low-input low-output farming systems have played an important role in shaping landscapes around the world. This is particularly the case in Europe where agriculture has been the primary land use for hundreds of years and the traditional farming systems have produced many types of semi-natural vegetation (Ratcliffe, 1984; Bignal and McCracken, 1996). It has been estimated that over 50% of Europe's most highly valued habitats occur within low-intensity farming systems (Bignal and McCracken, 1996) and these traditional systems are

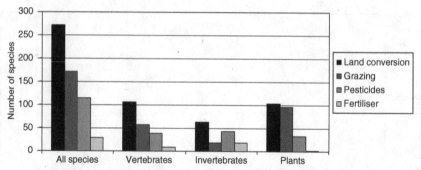

Figure 3.7 Number of native species threatened and endangered by agricultural activities in the United States. Note: land conversion refers to conversion from non-agricultural land use to cropland.
(Source: USDA, 1997)

essential to their continuation. Of the 663 plant and animal species listed as threatened or endangered in the United States, 380 are listed, at least in part, due to activities associated with agriculture, with the conversion of land to agricultural production being the primary threat (USDA, 1997) (Figure 3.7). In North America and some other regions, over the last 30 years, habitat loss due to conversion of land to agriculture has reduced wild species numbers more than any other human activity (McKenzie and Riley, 1995). Agriculture is now the dominant land use in North America. In the United States, over 50% of the land in the contiguous 48 states is allocated to cropland (23%) and grassland pasture and range (30%) (USDA, 2006a) while in the prairies of Western Canada, 93% of the land is allocated to agriculture (McRae *et al.*, 2000). Agricultural activities, from converting land to annual crop production, intensive grazing, and the use of inputs such as fertilisers and pesticides, and the highly specialised types of farming systems practised nowadays have all had a significant impact on both the landscape and wildlife.

The expansion of agricultural output from the 1950s, supported by both agricultural policy and advances in technology, has resulted in: (1) the loss of many natural and semi-natural habitats, (2) a decline in habitat quality, (3) fragmentation of surviving habitat patches and (4) simplification of landscape level diversity. As a consequence of the increasing level of inputs available to farmers, such as mechanisation and fertilisers, many lands that were unproductive, which could not be cultivated, were transformed into areas of intensive agricultural production. The expansion of agriculture onto previously marginal land and intensification of production on these lands (Table 3.1) was at the expense of many important semi-natural and natural habitats. In Europe, where little of the natural vegetation remains, many of the semi-natural habitats created by low-intensity farming systems such as heathland and species-rich grassland have declined enormously in area since the 1940s (Baldock, 1990). For example, in the United Kingdom the area of species-rich grassland is only 1% to 2% of permanent grassland area (Blackstock *et al.*, 1999). In North America, where agriculture has been a major land use for a much shorter period of time in comparison with Europe, the expansion of agricultural land use resulted in the loss of natural habitats, such as prairie and woodland. In the United States almost half of all wetlands in the 48 contiguous states have been drained since colonial settlement with nearly 85% of this loss attributable to agricultural use, with average net rates of conversion of approximately 330 000 to 360 000 ha yr^{-1} between 1885 and 1954 (Hansen, 2006). However, this rate of loss slowed significantly after this period with the introduction of conservation policy initiatives and smaller areas of wetlands that are relatively inexpensive to drain. In the Canadian prairies less than 1% of the native tall-grass prairie, 19% of the mixed-grass prairie and 16% of the aspen parkland remain of these original native plant communities (McRae *et al.*, 2000). In Europe hedgerows and ponds were removed by farmers mainly for economic reasons to enlarge fields and improve efficiency. In the United Kingdom, it has been estimated that 28% of hedgerows were lost between 1947 and 1974; a total of 225 000 km of hedgerow were removed (Pollard *et al.*, 1974).

The rarest and most vulnerable habitats and landscapes within Europe are now protected as any impacts of any land-use change on the environment have to be considered following the extension of Environmental Impact Assessment (EIA) regulations in 2002. The Environmental Impact Assessment Directive (85/337/EEC) was originally introduced in 1985 to reduce the environmental impacts of a broad range of development projects. The continued loss of habitats from increasing agricultural intensification resulted in the regulations (2003/35/EC) being extended to cover uncultivated land and semi-natural areas, including unimproved grassland, heath, moorland, scrub and wetlands. To reduce the risk of these habitats being destroyed the impact of any changes in land use, cultivation, fertiliser and lime application, drainage, flood defence and increased stocking rates should be assessed. Several farmers have been prosecuted successfully for failing to obtain permission and have been required to reinstate land to its previous condition. Other countries have developed endangered species legislation and a range of agri-environmental and environmental policies to address the impact of agriculture and agriculture development on wildlife species (see Chapter 4).

The integrated low-intensity farming system that included both crop and livestock production produced a diverse small-scale patchwork of different habitats within the agricultural landscape. The close proximity of these different habitats, such as woodland, along with the occurrence of cereals and grass within the landscape plus the diversity of crops grown provided important habitats for many species of mammals, birds, invertebrates and plants. Therefore, any change in the farming system or agricultural management had immense potential to influence the distribution and abundance of many species. Though it has been well documented that the decline of many important species has coincided with the changes in farming practices over the last 60 years (Fuller *et al.* 1995; Rich and Woodruff, 1996), many factors have been involved in the transformation from integrated low-intensity farming systems to the highly specialised types of farming system seen today. These include the separation of arable and livestock production (polarisation) and the associated changes in land use and cropping patterns together with the increasing level of inputs (intensification), such as mechanisation and fertilisers. The contribution of individual changes in agricultural management to the decline in farmland biodiversity and many species of mammals, birds, invertebrates and plants is difficult to determine as the many factors associated with the modernisation of agricultural production have had a part to play and the interactions between the factors are complex.

As well as decreasing the area of many natural and semi-natural habitats modern farming practices have also reduced the diversity of existing agricultural landscapes (Robinson and Sutherland, 2002). First of all, the structure and scale of the agricultural landscape has changed greatly following the polarisation of arable and livestock production. In many rural areas either intensive arable or livestock production now dominates the landscape. Consequently the diverse small-scale mixture of habitats, cereals and grassland, which was once widespread, has declined enormously (see Chapter 9 for the importance of landscape scale). Secondly, the influence of price incentives on the types of crops grown has led to huge changes in cropping patterns (Figure 3.3). Thirdly, mechanisation has allowed the simplification and concentration of farming operations into a shortened period of time. Pre-mechanisation, farming operations took place over a much longer period of time. For example, not all cereal stubbles could be ploughed immediately with the use of animals and were therefore staggered throughout the post-harvest period. Following mechanisation, stubbles could be ploughed in much more quickly following harvest. Lastly, the increasing level of inputs (fertilisers, pesticides and herbicides) has allowed the continuous production of one crop on the same piece of land. All these changes in agricultural practices have decreased the availability of suitable habitat for many species of mammals, birds, invertebrates and plants and subsequently reduced the diversity of farmland.

Over many years, a unique flora of annual herbaceous plants or 'arable weeds' had developed in association with arable cultivation. In response to the intensification and modernisation of agricultural practices there has been a major decline in these annual plants and many are now considered rare. In the United Kingdom, 18% of the plant species listed under the Biodiversity Action Plan are considered weed species of arable land. The decline of the arable weed flora is the result of several changes in arable cultivation. The practice of seed cleaning, the increasing use of herbicides and fertilisers, and changes in the variation in the timing of cultivation have all had a major impact on these 'arable weed' species (Ratcliffe, 1984; Wilson and King, 2003). The introduction of threshing machines enabled the removal of many of the contaminants of cereal crops with increasing efficiency resulting in a reduction in the seeds of 'arable weed' species. Prior to the use of synthetic herbicides, weeds were controlled by rotating the crops and then by either hoeing or the use of chemicals such as ferrous sulphate. The increasing number and selectivity of herbicides available to the farmers allowed them to control 'arable weed' species with much greater effect. In addition, the increased use of herbicide-tolerant genetically modified crops has enabled farmers to more effectively control weed species in annual crops. As a result, many broad-leaved species, a valuable food resource for many species of birds, have declined.

As annual plants, many arable weed species require a regular disturbance of the soil to germinate. More species of arable weeds are adapted to spring-sown crops. Changing the timing of cultivation can have major impacts on the arable weed species. Changing from spring to winter cereals has been shown to reduce the density and change the composition of arable weed species (Hald, 1999). Plants of arable habitats are not the only group of species to have seen a decline. Plants of grassland and heathland habitats have also declined, with the increased use of fertilisers and herbicides the primary cause (Rich and Woodruff, 1996). Many unimproved grasslands have been improved either by reseeding with productive grasses such as *Lolium perenne* or though the addition of fertilisers. It is very difficult to combine both the agricultural and environmental objectives in grassland management as the increase in productivity invariably leads to the loss of plant species within the sward and a decline in diversity (see Figure 1.9 in Chapter 1). As a consequence of the increasing level of inputs available to farmers, the agricultural landscape has become increasingly uniform in nature. However, theoretically diversity and productivity should be compatible; the relationship between the two remains controversial and agriculture is still a long way from being able to combine them.

One of the impacts that has given rise to most public concern that can be attributed to these changes in agricultural practices is the observed decline in farmland bird species (Chamberlain *et al.*, 2000). In the United Kingdom, the population of farmland birds declined by 45% between 1970 and 1999, although

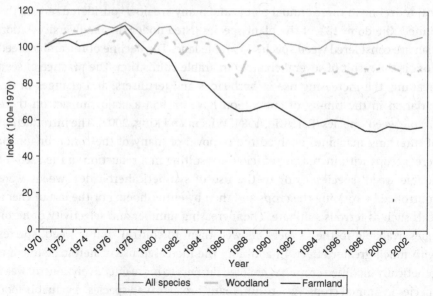

Figure 3.8 Wild bird population index in the United Kingdom 1970–2004.
(Source: Defra, www.statistics.defra.gov.uk/esg/default.asp/, last accessed 2006)

there has been a slight recovery in numbers since (Figure 3.8). In North America 70% of grassland bird species have been in decline for the last 40 years. The decline in farmland bird numbers can be attributed to many different factors: the loss of habitat including field boundaries such as hedgerows, changes in grassland structure through grazing and fertilisation, the increasing use of pesticides and herbicides and the subsequent decrease in food availability, the switch from hay to silage making and the resultant uniform grassland and the reduced availability of winter feed resulting from the change from spring- to winter-sown cereals (Fuller *et al.*, 1995; Vickery *et al.*, 2001; McCraken and Tallowin, 2004).

The environmental changes that have taken place in response to modern farming practices have been very rapid. The loss and fragmentation of habitat, the reduced diversity of crops and the concentration of farming operations have radically changed the appearance of the countryside. The reduction in both spatial and temporal diversity in the agricultural landscape has been huge and as a consequence few species have been able to adapt. It is increasingly recognised that low-intensity farming systems are of critical importance in the conservation of many habitats and the species they support. However, whether there is a future role for these low-intensity farming systems and whether this role in conserving biodiversity and valuable habitats can be reconciled with modern agricultural practices is still open to question.

Summary

Over the last 60 years the unprecedented increase in food production, driven by advances in technology and supported by agricultural policy, has come at considerable cost to the natural environment and farmland biodiversity. In terms of monetary value the costs are huge, with the emissions of gases, declines in population of wildlife and contamination of water by pesticides being the greatest costs. It has been estimated that the annual cost of UK agriculture is over £200 ha^{-1} (Pretty et al., 2000). The highly specialised intensive types of farming system seen today have replaced the traditional low-intensity farming systems, which were so important to conservation. It is now recognised that agriculture and conservation are interdependent, as the conservation of many species and habitats depends on agricultural management and agriculture depends on provision of ecosystem services such as requiring insect pollinators and predators. In addition, agriculture depends on biodiversity to be used in the development or adaptation of new varieties of plants to keep pace with new plant diseases, insect pests and changing climatic conditions. It is also acknowledged that the pressures exerted by modern farming on the natural environment and wildlife are likely to continue so long as the human population continues to increase. However, in most developed countries the majority of land within agricultural landscapes is privately owned and managed for private gain. Therefore, the sustainability of these systems and the long-term provision of ecological goods and services will need to be within a system of agricultural management not to the exclusion of agriculture.

4

Principles behind agri-environment schemes

Introduction

Agri-environment schemes have changed over time and differ, often dramatically, between countries and regions. The schemes differ in the rules, mechanisms of delivery and objectives, although the general principles that the schemes are based on are typically very similar. This chapter will at first focus on describing the socioeconomic motivation for implementing agri-environmental measures by developing a simple economic model. Following this the discussion will focus on highlighting the general agri-environmental mechanisms that have emerged over time and in different jurisdictions. We will also explore the universal problems of setting and achieving regional and national conservation objectives, the equitable and efficient allocation of resources, the selection of delivery mechanisms including targeting approaches and the acceptance and uptake of measures by an agricultural industry that is, often, extremely heterogeneous.

The motivation for agri-environmental policy

An important reason for implementing agri-environmental schemes is based on the economic concepts of externalities and market failure. Economic theory indicates that when there is a market failure the allocation of resources does not maximise the welfare of society. In the case of agri-environmental concerns, then, the presence of a market failure results in too little investment in the provision of ecological goods and services from society's perspective. In other words, agricultural management decisions result in too few environmental goods (e.g. biodiversity) and/or too many environmental bads (e.g. nutrient pollution).

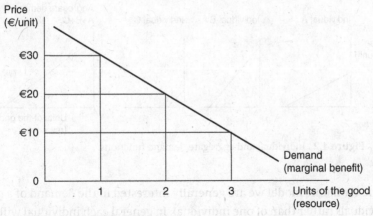

Figure 4.1 Basic demand function.

The basic market failure that is the focus of many agri-environmental schemes will now be described using a basic economic model. The economic model represents both the consumer and the producer side of the market and their interactions within the market. For the present discussion we will assume that the consumer represents society in general, and the producer represents farmers and land managers. Individual consumers ascribe value to goods and services and these values can be observed and measured, and are reflected in the amount consumers are willing to pay (in money, time, or other resources) to acquire that good or service. The willingness to pay for a good or service by a consumer (or group of consumers) can be represented by a demand function. The demand function shows how much of a good or service consumers demand at different price points (Figure 4.1). The demand function is often called the marginal benefit function in that it represents the benefit gained by consuming the next unit (marginal unit) of the good. The demand function in Figure 4.1 reflects that the individual is willing to pay €30 for the first unit of the good (implying that the value of the first unit is €30), €20 for the second unit and €10 for the third unit. In economics it is generally assumed that goods and services are infinitely divisible such that we can derive a smooth demand function (e.g. the demand function can represent the value of 2.5 units) rather than a step-shaped horizon. The downward slope of the demand function reflects the fundamental relationship of diminishing willingness to pay – as the number of units consumed increases, the willingness to pay for additional units of the good decreases. For example, a consumer would ascribe a large value to the first hectare of wildlife habitat preserved within a region but would ascribe a lower value to 1000th hectare. In addition, the total benefit received from consuming a given quantity of the good or service is represented by the area underneath the marginal benefit (demand) function.

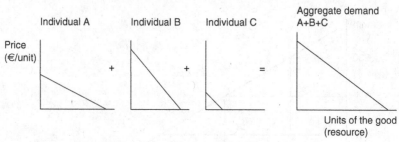

Figure 4.2 Individual and aggregate demand functions.

In the economic model we are generally interested in the demand of a group of individuals rather than of one individual. In general each individual will have a slightly different demand function. For example, one individual may be willing to pay €30 for the first unit of wildlife habitat while another individual may only be willing to pay €10 for the same unit of wildlife habitat. The willingness to pay for a good or service is influenced by characteristics of the individual including levels of income, individual beliefs and preferences and levels of knowledge about the good or service. The demand function for a group of individuals is simply the aggregation of the individual demand functions (Figure 4.2). This aggregate demand function is often viewed as the social demand function, reflecting society's demand for a particular good or service.

While the demand function represents the consumer agents operating in the economic market, the supply function represents the producer agents. It is generally true that the production or provision of anything will require the expenditure, or using up, of things that have economic value (to get outputs it is necessary to use inputs). For example the production of a tonne of wheat requires the expenditure of a wide range of inputs including out-of-pocket cash costs (e.g. labour, seed, fertiliser, fuel) as well as the costs of time and environmental costs (e.g. pollution, resource degradation). In the economic model it is generally assumed that the market prices of inputs (wages, fuel, materials) reflect the true cost of the inputs. Based on this, the primary relationship of interest is between the rate at which something is produced and the costs of that production. Once again we will consider the marginal cost, which is the cost associated with producing one more unit of the good in a period of time (Figure 4.3). From Figure 4.3 we can see that the first unit costs the producer €10 to produce while the marginal cost of the second unit is €20 and so on. Therefore, the function that represents these costs can be interpreted as a supply function since it reflects the quantity of the good the producer is willing to supply at different prices. The upward sloping nature of the supply function reflects a generic characteristic of production, that of increasing marginal cost.

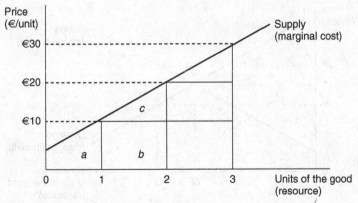

Figure 4.3 The marginal cost or supply function.

For example, an individual farm with a fixed quantity of land can only produce more wheat by increasing investment in inputs such as fertiliser, seed, irrigation water and labour. Once again it is assumed that output units are divisible enabling a smooth supply function rather than a step function. The total cost of production of a number of units of the goods is the total area under the supply function up to the marginal cost. For example, the total cost of producing two units of the good in Figure 4.3 is represented by area $(a+b+c)$.

The marginal cost curve function for each producer or firm is unique as determined by the technical and economic characteristics of the production process. Therefore, the height, shape and steepness of the marginal cost function reflect the quantities of inputs required to produce different levels of output. For example a farm located on level, highly fertile soil will be able to produce a tonne of wheat at a different cost than a farm with hilly, low-fertility soils. As with the demand function, where we are interested in the social demand, on the production side we will tend to be interested in the supply characteristics of a group of farms or an entire sector of the agricultural industry. As before, the aggregate supply function is determined by summing the individual supply functions.

Demand functions and supply functions are the components of the economic model that is used to determine how resources are to be allocated. The basic model assumes that the demand function (marginal benefit) captures all of the benefits received by all members of society associated with another unit of output, while the supply function (marginal cost) captures all of the costs imposed on all members of society by producing another unit of output. Therefore the curves are often referred to as social benefit and social cost curves respectively. Given these assumptions the socially efficient level of output is represented by price P^* and output quantity Q^*, and is defined as the output level

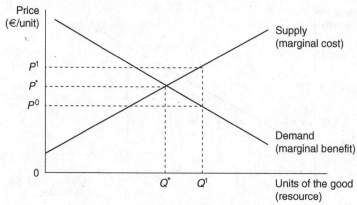

Figure 4.4 Basic economic model and social efficiency.

where marginal social benefits equal marginal social costs (supply = demand) (Figure 4.4). This price and output combination is considered efficient because the net benefits to society are maximised. For example, we can show that an additional unit of production beyond Q^*, for example Q^1, would impose greater marginal costs (represented by P^1) than the marginal benefit received (represented by P^0) (Figure 4.4). Therefore, the net benefits (total benefits minus total costs) at Q^1 would be smaller than at Q^*. Similarly, producing a unit less than Q^* would mean that positive net benefits (total benefits > total costs) are being forgone. In summary, the economic model determines the efficient allocation of resources whereby no other allocation would provide greater net benefits to society.

The model described dictates that if all assumptions hold then the exchanges between consumers and producers within a given market will result in an allocation of resources that is efficient and thereby maximises social net benefits or social welfare. However, as indicated earlier, one of the necessary assumptions is that the marginal benefit and marginal cost functions capture the full range of benefits and costs associated with the production and consumption of the good or service. In some situations there are a number of costs and benefits associated with the production or consumption of goods and services that are not captured by the producer or the consumer. These are called external costs or external benefits, or more generally, externalities. In the presence of externalities the marginal cost (marginal benefit) function will not reflect the true cost (benefit) associated with production of the market commodity and a market failure results.

External costs and external benefits are most often caused by the presence of public goods. Public goods are characterised as being non-excludable, meaning that once the good has been provided it is very difficult or impossible to prevent

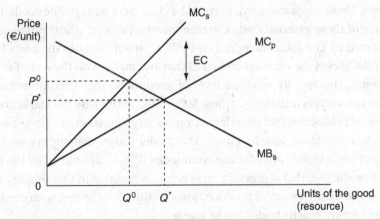

Figure 4.5 Basic economic model with market failure.

access to the good. Public goods can also exhibit a degree of non-rivalry, mean-ing that the use of the good or service by one individual does not diminish the amount available to another individual. For example, the benefits gained from biodiversity, once provided, are difficult to exclude people from and one per-son's use does not generally diminish the quantity or quality available for others. As a result, producers of biodiversity will find it difficult to obtain an economic reward through the market. In this case the goods and services provided by biodiversity would be considered an external benefit associated with certain agricultural management decisions. It is worthwhile to note that the characterisation of agriculture as a multifunctional industry implies that non-commodity goods and services, often public goods, are produced jointly with commodities.

Externalities are relatively common in production systems that use the natural environment as a source of production inputs, or provide environmental goods and services as joint products with market commodities, such as agricultural systems. For example, consider a wheat-producing farm that is releasing nutrient and pesticide pollution into a nearby stream. The manager of the farm will recognise the costs of purchasing and applying seed, fertiliser and pesticide inputs, and the costs of labour, equipment and buildings, which are captured by the private marginal cost function (MC_p) (Figure 4.5). In this example we will assume that there are no external benefits such that the marginal benefit func-tion (MB_s) reflects the full benefits associated with wheat production. Based on these functions the efficient level of production is Q^*. However, in this case there are a number of costs of production that are not captured by the farmer. These external costs include costs to downstream communities due to poor water quality, decreased stream biodiversity and decreased recreational fishing activities. These goods and services could be characterised as public goods. In the presence of these external costs the true marginal cost of wheat production is represented by the social marginal cost (MC_s), which includes the costs to the farmer (MC_p) plus the external costs (EC) that are imposed on the rest of society. As a result, the socially efficient level of production that maximises the net benefits to society is actually Q^0, where $MC_s = MB_s$. In this case the market results in a level of production that is too large from society's perspective ($Q^* > Q^0$), in that it does not maximise social welfare. The model shows that society would be willing to pay a higher price for the commodity ($P^0 > P^*$) as implied by the lower level of production that is efficient. In general, as revealed in this simple model, public goods (bads) are under-produced (overproduced) compared with goods and services that are readily traded in the market.

In general, public goods and the associated market failure are understood to be the primary reason for agriculture systems not providing socially efficient levels of ecological goods and services. However, other reasons have been identified to cause market failures in agricultural systems. For example, information failures result in inefficient allocations when agents (consumers and/ or producers) do not have full information on such aspects as relative costs or alternative available technologies. There is evidence that farmers do not have full information about the most efficient use of fertilisers, which results in nutrient run-off and pollution (Defra, 2002e). Government market failures occur when programmes are implemented, often with objectives such as encouraging production, that diminish or mask the environmental costs of management practices. For example, an input subsidy that decreases the cost of fertiliser will result in an increase in the use of fertiliser and possibly greater nutrient pollution than would be experienced without the input subsidy. This type of government intervention often exacerbates existing market failures. Finally, in the economic model it is assumed that producers make decisions to maximise profit or welfare. However many farmers may not make management decisions that are profit or welfare maximising and therefore will not lead to efficient resource allocation. This is due to such factors as 'path dependence', where previous decisions have precluded making future optimising decisions due to management skill, capital investments or age of manager.

General mechanisms of agri-environmental policy

Since the 1980s a wide range of policy measures have been developed to address environmental issues in agriculture. Agri-environmental problems have proven to be quite complex and very heterogeneous across a landscape and over time. In addition the agricultural industry is highly heterogeneous with farms, even within a region, having very different socioeconomic and biophysical characteristics. As a result a diverse range of agri-environmental policy measures have been developed and implemented. Generally, these measures are aimed at correcting or overcoming the market failure discussed above and thereby have the explicit or implicit objective of providing ecological goods and services at levels that are closer to a socially efficient level. The policy measures are developed specifically for the agricultural sector or as part of broader national or regional environmental programmes that influence the decisions of a number of sectors. The most common policy tools can be broadly categorised as economic instruments, command and control measures that enforce environmental beneficial management or restrict environmentally damaging management, and advisory and institutional measures that put in place information and institutions to facilitate environmentally beneficial decisions (Figure 4.6).

Economic instruments

Economic instruments are measures that change the costs and benefits and, therefore, the economic price signals driving the land-use decisions of farmers. Through this mechanism the policy can encourage positive environmental outcomes within agricultural systems. In general, economic instruments involve either a monetary transfer (e.g. through payments to farmers or taxes/charges imposed on farmers) or the creation of new economic markets (e.g. tradable rights or permits) (OECD, 2003b).

Payments

A common agri-environmental policy tool involves the offering of payments to farmers to encourage actions that provide ecological goods and services. The OECD (2003b) identifies the three main types of payment programmes as: (1) payments based on farming practices; (2) payments based on resource retirement; and (3) payments based on fixed farm assets.

Payments based on farming practices provide annual payments to farmers who adopt more environmentally beneficial management strategies. In essence these payments increase the likelihood that farmers will adopt the environmentally desirable practice by reducing the net cost of doing so (Claasen *et al.*, 2001). Greater payment levels are usually associated with greater environmental management expectations. However, participation in these programmes is voluntary and programmes that provide less than 100 per cent of adoption costs will be adopted only if the targeted management practices provide private economic

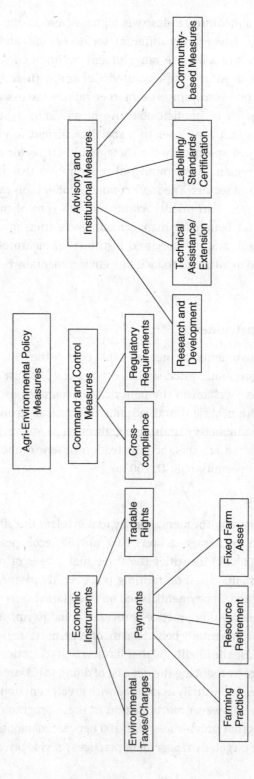

Figure 4.6 Agri-environmental policy measures. (Adapted from OECD, 2003b)

benefits. Increased adoption comes with greater payment rates but the programme can become expensive for taxpayers and in some cases may not be the least expensive way to achieve environmental objectives (Claasen *et al.*, 2001). Farming practice payments can be linked to the adoption of less input-intensive (e.g. decreased fertiliser or pesticide use) production strategies. A number of countries have also implemented payment programmes to encourage the adoption of nutrient and manure management. These payments focus on decreasing the external costs associated with, for example, nutrient pollution of surface and ground water or pesticide impacts on invertebrate biodiversity. Examples of this type of programme include European Union state initiatives supporting the conversion to organic production systems. Payments can also be tied to management practices that conserve, preserve or create biodiversity, wildlife habitat or landscape scenic amenities. This type of payment can be viewed as either decreasing the external costs (e.g. loss of biodiversity) or increasing the external benefits (e.g. habitat provided by a type of agricultural management) associated with agriculture. These payments are a means of public provision of public goods (ecological goods and services) through the government compensating farmers for this provision.

Payments based on resource retirement focus on providing incentive payments to remove land or other factors of production from management for agricultural commodity production, and thereby allocating the resources to ecological goods and services production. Land retirement programmes are an important component of agri-environmental programmes, particularly in the United States. For example, the US Conservation Reserve Program (CRP) provides rental payments to farmers to retire land from production for 10 to 15 years. Environmental objectives of these land retirement programmes include providing wildlife habitat, water and air quality improvements and carbon sequestration. Land retirement programmes are particularly appropriate for environmental benefits that increase with the length of time the land is idled (Claasen *et al.*, 2001). For example, wetland and wildlife habitat benefits are provided only after the ecosystem is well established, a process that may take a number of years. Land retirement programmes are also appropriate for taking environmentally risky land (e.g. highly erodible land) out of production. However, this type of programme is not appropriate at very large scales and can be very expensive. Another type of resource retirement programme focuses on providing compensation to farmers involved in environmentally damaging production to exit the industry. For example, pig farm buy-out schemes have been implemented in the Netherlands and Belgium to decrease manure surpluses (OECD, 2003b). These resource retirement programmes facilitate a form of public production of environmental public goods.

Payments based on fixed farm assets are aimed at offsetting the investment cost of adjusting the farm infrastructure and/or farm equipment to enable the adoption of more environmentally beneficial farming practices. Payments can be targeted at offsetting the costs of creating or improving animal waste management facilities; establishing or improving buffer strips adjacent to surface water sources; establishing shelter belts and windbreaks; maintaining, improving or creating hedges or stone walls; and the development of upland, wetland, riparian and aquatic habitat areas. A relatively common motivation for this type of programme is to provide assistance to farmers to help them meet the requirements of specific environmental regulations.

Environmental taxes and charges

Environmental taxes and charges include those policy measures that impose an additional cost on farm inputs or outputs that are a potential source of environmental damage. In the context of the economic model taxes and charges serve to 'internalise' the external costs associated with an input or management practice. As such, this policy measure assumes that farmers should bear the costs of complying with regulations addressing pollution or environmental degradation. However, the OECD (2003b) reports that taxes and charges are rare in agriculture compared with other sectors. This may be due to the fact that, for example, pollution from agriculture is heterogeneous and highly dispersed and not point-source as found with other industry. In addition, taxes and charges are sometimes viewed to be in conflict with farmers' property rights and will have a negative impact on farmer income making them politically unattractive. Nonetheless, taxes and charges have been imposed on such inputs as pesticides, fertiliser and water with the objective of decreasing their use.

Tradable rights

Tradable rights, or quotas, involve the establishment of environmental permits, restrictions, maximum rights or minimum obligations that are tradable or transferable. These rights or quotas are granted to farmers who can transfer them, at a price, to those who value them most highly, facilitating economic efficiency. This approach attempts to address the market failure by creating an economic market for environmental goods and services that are not normally traded in the market. The application of tradable rights in agriculture is relatively rare, probably due to the high cost of setting up and monitoring the system. However, examples of this policy measure include systems of manure production quotas, water extraction rights for irrigation and wetland mitigation credits. In addition, as a component of greenhouse-gas emission reduction initiatives, a carbon-trading system is being investigated whereby farmers will be able to sell soil carbon sequestration credits to industrial greenhouse-gas emitters (Thomassin, 2003).

Command and control measures

Command and control measures use compulsory restrictions on the choices of farmers such that they are required to comply with the rules and restrictions imposed or they will face penalties including fines and withdrawal of financial support. Command and control measures can be categorised broadly as regulatory requirements or cross-compliance measures.

Regulatory requirements

Regulatory requirements are compulsory standards or requirements on farmers to achieve specific levels of environmental quality and include environmental restrictions, bans, permit requirements, maximum rights or minimum obligations. Failure to meet these requirements is enforced using courts, police or fines. Regulatory requirements are viewed as being less flexible than economic instruments since they do not allow farmers to decide, for themselves and their farm, the most appropriate way to meet the environmental objective. However, regulatory requirements do minimise risk and uncertainty by providing a specific goal and are often considered appropriate for acute environmental problems and/or where environmental degradation may be irreversible. Claasen *et al.* (2001) state that regulatory requirements can be the most effective of all policy tools in effecting changes to improve environmental quality, assuming that the regulations are adequately enforced. As a result, these measures have been used extensively in the agriculture sector to address a range of environmental costs. Regulations and standards have been used to decrease pollution by restricting the application and discharge of nutrient, controlling the location and time of pesticide application and requiring the establishment of buffer strips (OECD, 2003b). Regulations have also been used to control natural resource use such as limits or caps on surface and ground water extraction, on soil management to limit erosion and salinity, and to meet national and international biodiversity and wildlife population objectives on agricultural land.

Cross-compliance mechanisms

Cross-compliance in the context of agri-environmental policy involves the required adoption of environmentally beneficial farming practices or attaining levels of environmental performance before farmers can receive agricultural support payments. A number of countries have made support payments based on output, factors of production or income conditional on farmers respecting certain environmental constraints or achieving particular environmental outcomes (OECD, 2003b). The motivation for cross-compliance is the belief that if farmers are being supported with public money then they should be responsible for providing some level of environmental quality. Examples of cross-compliance include making farmers ineligible for farm programme payments if they do not employ conservation management on highly erodible land, convert wetlands to agricultural production or generally do not comply with environmental standards and farm-management practice requirements. In the United Kingdom farmers are only eligible for support payments under the Common Agricultural Policy of the European Union (CAP) Single Payment Scheme if they meet a series of specific environmental management and environmental condition requirements. Under the European CAP the 'minimum environmental standards' that comprise the cross-compliance requirements are considered to represent society's baseline environmental expectations of agriculture. Environmental commitments above this reference level may qualify for agri-environmental payments (European Commission, 2006). In a number of applications cross-compliance indirectly attempts to correct the distorting influence of non-agri-environmental government agricultural support programmes.

Advisory and institutional measures

Many countries include advisory and institutional measures as a component of their agri-environmental policy framework. Institutional measures involve the government providing the institutional foundation to facilitate farmers participating in community and collective projects to address local or regional environmental issues. Governments invest in advisory measures to improve information flows to farmers and to consumers to promote and mediate the meeting of environmental objectives.

Research and development

Research into the interactions between agriculture and the environment has become an important component of agri-environmental policy. In many countries the research is focused on establishing best or beneficial management practices (BMPs), which are management practices that either minimise the environmental impact of production systems or are environmentally beneficial. The research on BMPs is then used to encourage farmer adoption through on-farm technical assistance or to inform the development of appropriate regulations or other policy measures. In addition, a number of countries have invested in the development of agri-environmental indicators that will be used to inform farmers and policy makers of trends in agri-environmental change. In recent years spending on agri-environmental research and development as a proportion of total agricultural research has increased in a number of countries (OECD, 2003b). The role of research and development is often to address information gaps that contribute to farm management decisions that cause environmental external costs. By providing this research information to farmers it is hoped that the agriculture industry will provide more ecological goods and services to society.

Technical assistance/extension

A complementary initiative to the research and development component of agri-environmental policy is the provision of technical assistance and extension. The purpose of these measures is to provide farmers with on-farm information and technical assistance to adopt and implement environmentally beneficial management practices (see previous section). These initiatives serve to inform farmers of the environmental issues specific to their farm and to induce voluntary changes in management practices to provide ecological goods and services. It should be noted that these management practices will be more readily adopted when there are private benefits to the producers, including decreased production costs, soil productivity enhancement and improvements in water quality. In fact, a disadvantage of this measure is that it is completely voluntary and effectiveness is dependent on adoption as determined by the private benefits provided (Ribaudo, 1997). Examples of this approach include the provision of technical assistance with the planning and implementation of management strategies to conserve soil, improve water quality, manage nutrients and manure and provide wildlife habitat. Some countries are encouraging or requiring farmers to complete environmental farm plans, which are often quite comprehensive evaluations of farm-specific environmental risks, to assist in the targeting of technical assistance.

Labelling/standards/certification

Governments have helped to establish the institutions, standards and certification bodies necessary for voluntary environmental or eco-labelling initiatives. These eco-labels help consumers distinguish products produced using environmentally beneficial management practices. The most common labelling initiative identifies organically produced food. However, labelling schemes have been developed for a wide range of agricultural products and production methods including agricultural products produced using integrated pest management, management methods that meet specific watershed conservation objectives or that provide bird habitat (Eco-labels website, 2006). Labels enable consumers to identify agricultural products that provide desired environmental attributes and at the same time enable farmers to receive an economic premium for products produced that provide desirable ecological goods and services. In this way labels can internalise the external benefit identified in the economic model. However, labelling will only be effective where private gains from participation can be captured in the market, and in some cases it will be difficult to link programme participation to measurable environmental benefits (Claasen *et al.*, 2001).

Community-based measures

Community-based measures are community-based groups, often government supported, that implement collective projects to improve environmental quality in agricultural landscapes. The government supports these group initiatives with research, planning, technical assistance and extension. The objective of these measures is to mobilise and motivate citizens to take on greater responsibility for addressing local or regional environmental issues (OECD, 2003b). Community-based measures focus on improving the flow of information between community members, facilitate the application of local knowledge and use peer pressure to attain results. In addition the approach can enable a landscape or watershed perspective to agri-environmental management. Examples of these measures have been used to increase the adoption of sustainable agriculture measures including land conservation, water conservation and vegetation management practices.

Achieving agri-environmental policy objectives

Given the wide range of policy measures and the many associated agri-environmental objectives it may be difficult to determine whether these objectives are being met. Since agri-environmental policy is a component of more traditional agricultural policy, which includes a range of programmes serving objectives from farm income support to environmental conservation to rural development, conflicts among objectives are inevitable. For example Europe's CAP objectives include 'helping agriculture fulfil its multifunctional role in society: producing safe and healthy food, contributing to sustainable development of rural areas, and protecting and enhancing the status of the farmed environment and its biodiversity' (European Commission, 2003d). With respect to the agri-environment, the agri-environmental measures are developed to address at least one of two broad objectives: (1) reducing environmental risks associated with farming (decreasing external costs); and (2) preserving nature, native and cultivated landscapes (increasing external benefits). At the farm or local level the agri-environmental measures have more specific objectives such as conserving a target area of wetlands or decreasing nutrient run-off to a target level. These more specific objectives may be complementary or conflicting and it may not be possible to achieve multiple objectives with a single policy measure. For example a policy aimed at decreasing pesticide pollution may result in increases in soil erosion while at the same time having a positive impact on biodiversity. Therefore setting objectives and understanding when this hierarchy of objectives is met is a non-trivial component of agri-environmental policy.

To help monitor environmental trends and to evaluate whether agri-environmental policies are meeting environmental objectives a number of governments have developed agri-environmental indicators. To effectively inform policy agri-environmental indicators must be:

- Policy relevant – address the key environmental issues faced by governments and other stakeholders in the agriculture sector.
- Analytically sound – based on sound science, but recognising that their development is iterative.
- Measurable – feasible in terms of current or planned data availability and cost-effective in terms of data collection.
- Easy to interpret – should communicate essential information to policy makers and the wider public in a way that is unambiguous and easy to understand (OECD, 2001b).

Agri-environmental indicators that meet these criteria have been identified as important tools to help governments understand the type of agri-environmental policy required, as well as whether the existing agri-environmental policy is meeting its stated objectives in an effective and efficient manner. A common framework for agri-environmental indicator development has been identified by the OECD (2001b):

- Driving force indicators – focus on the causes of change in environmental conditions in agriculture, such as changes in farm management practices and the use of farm inputs.
- State indicators – highlighting the effects of agriculture on the environment, for example, impacts on soil (e.g. soil carbon stock), water (e.g. phosphorus loads in surface water) and biodiversity (e.g. species richness).

- Response indicators – focus on the actions taken to respond to the changes in the state of the environment, such as variations in agri-environmental research expenditure.

This framework enables agri-environmental indicators to capture trends in agricultural industry that will influence the environment as well as the trends in environmental quality.

While indicators are being developed as an important component of a number of countries' agri-environmental policy frameworks, there are some characteristic shortcomings that should be considered. Agri-environmental indicators may not be effective in cases where the environmental benefits provided by a policy measure are difficult and/or very expensive to measure. For example, biodiversity conservation as a policy objective will in some cases be very difficult to measure in a meaningful way across a landscape. Site-specific inventories may be possible but how that translates to landscape-level biodiversity may be uncertain. Due to the cost of developing very fine-scale indicators the agri-environmental indicators developed often use regional data. Indicators based on regional data will be effective for certain agri-environmental concerns but may not be for other issues. Finally, agri-environmental indicators may have limited ability to measure policy success when the link between management change and environmental change is difficult to identify. In some cases the environmental improvement may be spatially and/or temporally separated from the change in agricultural management. For example the benefits from decreased nutrient pollution may be experienced some distance downstream from the farms that have adopted the pollution-reducing management practices. Further, conserved wildlife habitat, such as wetlands, may not provide the desired environmental benefits for many months or years after conservation due to the time required for that ecological function to be restored.

Targeting of agri-environmental policy

The effectiveness (meeting policy objectives) and the efficiency (greatest benefits at lowest cost) of policy measures will be significantly influenced by how the measures are deployed spatially and temporally. This is particularly true for agri-environmental policy given the size of the agri-environmental budgets, the range of environmental problems to be addressed, and the geographical size and heterogeneity of the area that often needs to be impacted. For example, to meet specific environmental objectives policy delivery should consider such factors as the economic productivity of land, the economic contribution of agriculture to rural communities, the vulnerability to natural hazards, the contribution to biodiversity and wildlife stocks, the contribution to non-point agricultural pollution and water quality and quantity impacts. As a result, policy targeting is receiving increasing attention in the development of agri-environmental policy frameworks.

In the absence of a process of designating or targeting funds to conservation priority areas, it is possible that the environmental benefits will be less, due to funds not being targeted to specific geographic areas, and the environmental effects of conservation practice implementation may be diluted by scattering funding across a broader area (Carpentier et al., 1998). For example, an approach that is often used to deliver agri-environmental schemes is to provide a fixed payment (per hectare) to all farmers who adopt a given management prescription. Inherent in this programme is a cost-targeting approach such that the only farmers who will adopt the management prescription will be those whose costs of adoption (or compliance costs) are less than or equal to the fixed payment. The pattern of adoption, and therefore the pattern of environmental benefits, from this approach will not necessarily provide the greatest aggregate benefits for a given agri-environment budget. For example, consider an agri-environmental programme that encourages the establishment of riparian buffer zones to meet surface water quality objectives. A fixed payment (cost targeting) will result in riparian buffers being established on land where it is least expensive for the farmer to do so, which will be the least productive agricultural land. However, if the most productive agricultural land, and therefore the most expensive to convert to riparian buffer, is the source of the majority of the pollution entering the surface water, then the agri-environmental programme will not have met its objectives. In this example, an approach that more specifically targets the land that is releasing the greatest quantity of pollutants may be more effective and more efficient.

An example of modifying the targeting of an agri-environmental programme has emerged from the CRP, a land set-aside programme in the United States. The CRP was initially established to meet supply management and soil erosion reduction objectives and was delivered to farmers who were willing to set-aside land for a set land rental rate. However, after approximately ten years the programme was modified to target land that could provide greater environmental benefits using the Environmental Benefit Index (EBI). The EBI is made up of a number of factors that account for environmental benefits (e.g. wildlife habitat, water quality, soil erosion, air quality) and contract costs. Some environmental factors are given more points (e.g. wildlife habitat, water quality) than others (e.g. air quality), because their non-market benefits are thought to be larger. The scoring of points for each EBI factor for each field is based on soil type, location, county population and the proposed CRP cover type. These points serve as a proxy for the relative value of the field's potential environmental impact. An evaluation study showed that adopting environmental targeting provided a $370 million per year increase in CRP benefits with programme costs unchanged (Feather *et al.*, 1999).

Cumulative effects and targeting

A number of environmental functions may exhibit cumulative or threshold effects. Such an effect exists when a significant environmental improvement can be achieved only after conservation efforts reach a certain threshold. For example, certain wetland functions, such as habitat for particular wildlife species, may require a threshold level of wetland concentration or wetland quality in a given area to be sustainable. If wetland habitat falls below this threshold the wildlife population is no longer viable. From a policy perspective, ignoring cumulative effects may cause conservation funds to be overly dispersed geographically and, as a result, produce the minimum environmental benefit when the conservation budget is small (Wu and Boggess, 1999). The presence of cumulative effects will mean that to meet particular environmental objectives, it will probably be suboptimal to distribute conservation expenditures across a landscape. It may be better to focus conservation activity in a specific area in order to ensure the threshold is met to provide the targeted benefits. Further, once this threshold has been met additional conservation in that area may no longer be necessary and policy should be redirected to another area. This points to the importance of managing and targeting conservation policy in ways that recognise the broader landscape or watershed and reflect the complexity of the environmental and ecological system.

Summary

The fundamental motivation for agri-environmental policy is to maintain or enhance social well-being by addressing environmental concerns that are caused by market failures in the agriculture industry. The economic model shows that many ecological goods and services have public good characteristics such that farmers will not have an incentive to provide them to society, resulting in an inefficient allocation of resources from society's perspective. The function of most agri-environmental policy measures, then, is to internalise the external costs and/or external benefits through economic instruments, command and control or advisory and institutional measures, such that farmers make management decisions that are more environmentally beneficial. The appropriateness of a particular policy measure is determined by the socioeconomic and biophysical characteristics of the farms and the regions, the preferences of society, political priorities as well as the existing policy environment.

5

Farm conservation planning

Introduction

Although this is predominantly a science book, the art of producing a
farm conservation plan requires an understanding of many diverse topics
including social skills, economics, history, geography, hydrology, not to mention agriculture and conservation ecology. Needless to say, no one individual
can be an expert in all these elements, so this chapter simply describes the main
elements that are needed in producing a general farm conservation plan that
would meet the requirements of a typical agri-environment scheme. Most
Organisation for Economic Co-operation and Development (OECD) countries
have developed farm conservation planning tools including Environmental
Farm Plans in Canada and New Zealand, Conservation Planning initiatives in
the United States, Farm Conservation Plans across Europe and Planning initiatives associated with the Land Care Program in Australia. While each conservation planning initiative is somewhat unique, all include general objectives of
identifying environmental risks and conservation opportunities of the target
farm. This chapter avoids covering in depth the quirks and foibles of any
particular scheme, because although all schemes appear to have these idiosyn-
crasies and although they are of some interest and often highly amusing, they
tend to be parochial and ephemeral. This chapter is designed as a practical
guide, identifying and ordering the main elements that should be considered
when producing a farm conservation plan with a particular focus on the UK
context. Some of these aspects may be intuitively obvious, but many issues are
easy to overlook unless you are a veteran of farm conservation planning.
Hopefully this chapter is a workable guide for those who have never produced
a farm conservation plan that covers all the basic elements required and indicates areas where specialist advice should be sought. For the more experienced

farm-conservationist, this chapter covers how to produce a workable agreement that benefits both the farmer and the environment.

First contact with the farmer

It is important to remember that agri-environment schemes are not compulsory, although cross-compliance, modulation, economics and various farm assurance schemes are effectively eroding this. Therefore, unless a farmer finds a proposed conservation plan acceptable, financially viable and agriculturally practical there will be no agreement, no environmental benefits and no support payments. The first meeting or phone conversation with a farmer is therefore crucial. A firm handshake and the ability to make small talk about the weather or the current state of farming etc. will be invaluable. Unless the farmer can see some benefit from the scheme you are both wasting your time. For historical and political reasons, different schemes may be operated by different organisations, which are likely to be perceived differently by farmers, and different farmers may contact different organisations to help produce their farm plans. The organisation you represent is likely to affect the initial impression the farmer forms of you. If you are a bearded, English, sandal-wearing ecologist dealing with a Welsh-speaking hill farmer, it may help if you are employed by a respected firm of agricultural consultants rather than by a conservation organisation. Whoever you represent it is imperative that you develop a working relationship with the farmer and you respect the fact that they are not unreasonably in the business of making money, and not growing wild flowers or looking after birds unless the two things are compatible. To reinforce this impression, farmers will often spell out this fact to farm conservation officers at their first meeting; subsequently they may reveal this is only partly true, as they delight in being shown something of great conservation value on their farm. Farmers also tend not to be big fans of administration and rule books. Unfortunately agri-environment schemes usually come with a long list of terms and conditions. Although they are important, it is best not to dwell on them in your preliminary discussions, except perhaps to make light of them. During your first contact with the farmer it is imperative that you determine what he/she wants from the process and discuss if his/her farming system is compatible with the particular scheme. If not, is he/she willing and able to embrace the changes required to become compatible?

Regional language

When dealing with farmers in a new area it is wise to be aware of variation in regional language; this is more marked in the countryside than between mobile urban communities. It frequently means replacing one word with another. But sometimes it can be more complex as words change their meanings. For example in England a dyke is a ditch or small stream, whereas in Scotland the same word is used for a dry-stone wall, and in the United States it has a different meaning again. North of the border the poisonous agricultural weed of grasslands *Senecio jacobaea* is called Tansy, but in England this name refers to the less common plant *Tancetum vulgare*, which is often a garden escapee found in waste ground.

Farm conservation/environmental audits

Virtually all agri-environment agreements require environmental audits to be carried out. They may vary in their complexity, depth and in their requirements for maps and background descriptions of the farming system, the wider landscape and locally important habitats and species. But if you are going to produce a plan to manage the conservation value of a farm, firstly, you must know what is present to be managed, what condition it is in and also what might be reasonably encouraged to inhabit the area. In addition, you will need to determine if there are currently any environmental issues of concern, e.g. poor waste management or other pollution problems; if not actually illegal, typically these must be corrected as a prerequisite to applying to any scheme.

Audit maps

You may think that obtaining a suitable map to produce a farm con-servation plan is straightforward – but this may not be the case! The quality and scale of map required is likely to vary between schemes; sometimes what is acceptable may even vary between administrators within schemes, so before you start establish what format of map is needed. Typically a 1:10 000 scale is ideal, but this may be problematic with very small hobby farms and with large hill-farms. Most farm conservation maps are still produced by field officers with sets of coloured pencils, but increasingly computerised maps linked to Geographic Information System (GIS) databases are being used. If these have been produced for agricultural purposes they may lack sufficient detail of non-farmed areas. On the plus side, they can be a great advantage when finding field codes, and calculating areas of habitats and lengths of field-margins.

Obtaining maps where the field boundaries are consistent between agricul-tural administrative maps, official government-approved maps, conservation agency maps and maps of historic features can be a real challenge. So, leave plenty of time before arranging a farm visit to obtain at least one, and ideally all, of the above. Even then you must be aware that field boundaries are likely to

have changed and you will inevitably spend lots of time redrawing them. Agri-
environment schemes have added to this problem. In the past a farm boundary
may have officially followed a meandering stream, but, for ease of fencing, field
boundaries have cut backwards and forwards across the water-margin. This is
fine until you want to exclude stock or associate payments to lengths of streams.

Schemes vary in their requirements to produce environmental audits for all
parcels of land owned and/or managed by a particular farm-business. This may
discourage a business that owns several intensive farms, but is only interested in
entering one of its more extensive holdings into a scheme. Even if these intensive
areas are not actively managed within a scheme, the auditing process will place
restrictions on them, preventing further habitat loss. Conversely, sometimes
when different schemes operate in different areas, occasionally separate audits
and agreements may be possible on different areas of the same farm and this may
encourage farmers into schemes for two sets of payments. Typically, if different
farm managers can be identified within a business then different audits and plans
can be drawn up for the separate farms. Be aware not to miss outlying individual
fields, and remember to ask about additional seasonally rented land, because this
may also need to be included in the audit. One last complication to think about at
this stage is different land designations for support, because within some
schemes different options may operate in land registered as marginal or upland.
It is best to be aware of this before doing the audit on the ground.

Preparing to carry out an environmental audit

Before investing time and effort in producing a farm conservation
audit, you may need some commitment from the farmer that he/she is genuinely interested in the scheme. This may be something of a catch 22, because
how can he/she decide unless he/she has some indication of what might be
involved in terms of management changes and financial rewards? Unless you
are already familiar with the farm, there may be no escaping doing a quick
reconnoitre of the holding or talking over possible options over a farm map.
Ideally, conservation auditing should be carried out in spring or summer and
many schemes have seasonal application periods to encourage this. However, in
the real world, audits are sometimes carried out under less than ideal conditions. But who is to say that the conservation interest of a site is not the visiting
wintering birds or the nocturnal wildlife and audits are virtually never carried
out in more than one season.

Carrying out a conservation audit is a difficult job that must be done correctly
because it will become the basis of a legal agreement; as such, all steps should be
taken to ensure that it is factually correct. Before you embark, it will save time
and energy if you look over the maps in detail with the farmer. Identify all areas

of potential interest; for example, ask the farmer to identify all field-margins of conservation value, e.g. hedges and dry-stone walls. Then plan your route around the property. It will be embarrassing and expensive if you need to make a return visit just to check a small area of habitat. Take a pair of binoculars with you, it will save a lot of walking. Ideally, also take the farmer with you, he/she will obviously know the land much better than you, but it is also important to have him/her involved in the agreement. If he/she feels ownership of the plan, and understands what he/she has and how to manage it, the process is more likely to be successful. However, farmers are busy people and they may not have enough time to walk the entire farm with you; if this is the case at least try and get them to drive around the more interesting areas with you.

What needs auditing?

What is conservation value? There are no units of conservation worth or agreed definition of it. Agri-environment schemes vary in the detail of what should be audited, but agree on the main elements considered worthy of conservation. These are any natural or semi-natural habitats (wetlands, species-rich grasslands, heaths, woodlands and scrub, etc.), any area of water (still or flowing), hedges and dry-stone walls. The list is likely to include features of historic or landscape interest such as policy grasslands (with traditional farming practices). Buildings and animal breeds may appear as optional extras. All of these may require colour coding on an audit map and their total areas or lengths calculating. Sometimes these features may overlap on audit maps; for example, species-rich grasslands may occur over archaeological remains, and this may need clarification in the associated text.

Currently agri-environment schemes are primarily defined in agricultural, not conservation, terms. The pros and cons of this are discussed in Chapter 4. The implications of this for environmental auditing are that the features that require inclusion tend to be defined by their land-use rather than by their ecological definition/description. For example, two identical areas of species-rich grassland may be audited differently if one occurs on inbye land and the second is located the other side of the fence line, which marks open hill ground. This is clearly nonsensical in ecological or conservation terms. However, within an agri-environment scheme, where payments are associated with potential agricultural income forgone rather than conservation values, it is logical, because in agricultural terms marginal hill ground has less economic value than more productive inbye land. For this reason, agri-environment schemes may require many features to be audited differently depending on if they occur on hill ground or not; the list not only includes species-rich grasslands, but also wetlands, heathland and some types of woodland. All of these are more likely to be classified as rough grazing (an agricultural definition rather than an ecological one). Similarly, streams, hedges and dry-stone walls may not need recording if they are located on hill ground. Always check with the particular scheme rules. Being a competent field botanist helps with producing an environmental audit, but it is not essential. Your identification skills will improve with time. Until then it is a good idea to carry a plant identification book.

Fuzzy definitions and habitat mosaics

Legal clarifications of habitat definitions have been produced in the European Court in relation to the Habitats Directive but apparently not yet for agri-environment schemes. This is perhaps surprising as habitat auditing can be a rather subjective method of determining if support payments are available or not.

All of the features that require mapping and/or describing in an environmental audit can be said to have fuzzy definitions. For example how many species are required for a grassland to be described as species-rich? How wet does a wetland need to be? What is the minimum number of stones in a dry-stone wall before it becomes a pile of rocks, etc? These questions can be problematic in audit terms because it forms the basis of a legal document. If features are missed off an audit, this could be said to be attempted deception, because by doing so the features would escape being protected by the agreement. Conversely, erroneously including features in an audit could be seen as dishonestly trying to increase the assessment of conservation value in an attempt to enhance the likelihood of receiving funding. Therefore, some of these definitions have been discussed at length. Occasionally guidelines have been produced to help (see Box 5.1). However, the majority of such questions have been left remarkably vague and in practice a great deal of flexibility seems to be applied.

Trained ecologists may prefer grassland, wetland and woodland habitats to be defined using formal scientific descriptions of plant communities such as the British National Vegetation Classification (NVC) system (Rodwell, 1991a, b, 1992). Such systems tend to be overly complex for the purpose, and they are not necessarily easy for farmers or scheme administrators to understand. For example, the NVC contains 12 semi-natural neutral grassland types alone and these seem unlikely to replace the more general classification of unimproved grassland used by schemes. It may be worth listing the plant community types present within the audit text.

As a general principle, if any doubt exists about whether a feature should be covered in an environmental audit, it is best to include it. If the scheme allows for descriptive text in the audit, then be honest. There is nothing to lose; if a feature has survived this long, the farmer is unlikely to want to destroy it during the period of the agreement. Unfortunately, this is not always the case, and sometimes farmers will try and encourage the omission of features from audits

Box 5.1 Guidelines For Identifying Unimproved Grasslands

How to identify 'improved grassland'

1. Choose ten points at random within the area of land under consideration. These points should not include field-margins, headlands or obvious areas of different vegetation.
2. At each of the ten points examine the vegetation in a circle of 1 m diameter.
3. Look for the following four indicator species: rye-grass (*Lolium perenne*), timothy (*Phleum pratense*), cocksfoot (*Dactylis glomerata*) and white clover (*Trifolium repens*) and estimate how abundant they are.
4. If six or more of the ten circles are more than half covered by one of the four indicator species, then the field can be considered **improved grassland**.

How to identify 'unimproved grassland'

1. Choose ten points at random within the field as above.
2. At each of the ten points examine the vegetation in a circle of 1 m diameter, looking for the occurrence of any of the indicator species listed below.
3. If six or more of the ten circles contain five or more indicator species, then the field can be considered **unimproved grassland**.

List of indicator species of unimproved grassland

Grasses

Quaking grass	*Briza media*
Sheep's fescue	*Festuca ovina*
Crested hair-grass	*Koeleria macrantha*
Meadow oat-grass	*Avenula pratensis*
Upright brome	*Bromus erectus*
Tor grass	*Brachypodium pinnatum*
False oat-grass	*Arhenatherum elatius*
Yellow oat-grass	*Trisetum flavescens*
Wavy hair-grass	*Deschampsia flexuosa*
Matgrass	*Nardus stricta*
Sweet vernal grass	*Anthoxanthus odoratum*
Purple moor grass	*Molinia caerulea*

Sedges and Rushes

Sedges	*Carex* spp.
Wood rush	*Luzula* spp.
Cotton grass	*Eriophorum* spp.

Herbs with basal rosettes

Rough hawkbit	*Leontodon hispidus*
Autumn hawkbit	*Leontodon autumnalis*
Stemless thistle	*Circium acaule*
Cat's-ear	*Hypochoeris radicata*
Cowslip	*Primula veris*

Herbs with uniformly leafy stems, and basal leaves usually smaller

Bird's-foot trefoil	*Lotus corniculatus*
Squinancy wort	*Asperula cynamchica*
Red clover	*Trifolium pratense*
Meadow vetchling	*Lathyrus pratensis*
Heath bedstraw	*Galium saxatile*

Herbs with basal and stem leaves (usually smaller)

Bulbous buttercup	*Ranunculus bulbosus*
Common knapweed	*Centaurea nigra*
Salad burnet	*Sanguisorba minor*
Greater burnet	*Sanguisorba officinalis*
Drop-wort	*Filipendula vulgaris*
Meadowsweet	*Filipendula ulmaria*
Ox-eye daisy	*Leucanthemum vulgare*
Hogweed	*Heracleum sphondylium*
Common sorrel	*Rumex acetosa*
Wood crane's-bill	*Geranium sylvaticum*
Lady's-mantle	*Alchemilla glabra*
Cuckoo flower	*Cardamine pratensis*
Tormentil	*Potentilla erecta*
Small scabious	*Scabious columbaria*
Mouse-ear hawkweed	*Hieracium pilosella*
Pignut	*Conopodium majus*

Annual herbs

| Fairy flax | *Linum catharticum* |
| Yellow-rattle | *Rhinanthus minor* |

Mat-formers

Wild thyme	*Thymus praecox*
Common rock-rose	*Helianthemum nummularium*
Horseshoe vetch	*Hippocrepis comosa*

Subshrubs

Bilberry	*Vaccinium myrtillus*
Heather	*Calluna vulgaris*
Heaths	*Erica* spp.

This list includes species associated with a range of different soil types and therefore this method can be applied to many types of unimproved grassland.

The methodology was developed by and is reproduced with the kind permission of the Forestry Commission of the United Kingdom.

to give them opportunity to remove them. This is not compatible with scheme rules and risks jeopardising the agreement and may result in financial penalties. Similarly, farmers may destroy features (for example ploughing species-rich grasslands or draining wetlands) before entering into schemes. Again this is not acceptable. It is difficult to obtain precise information about how common this practice is, but a rough estimate is worryingly in the order of between five and ten per cent of applicants.

Another potential problem that may be faced when drawing up an environmental audit is the habitat mosaic. This commonly occurs when patches of wetland, species-rich grasslands, heathland, bracken and scrub occur within single fields. Such mosaics are difficult to map accurately by eye. Remote sensing may increasingly be the solution. The issue is less of a problem when the mosaic is a mixture of semi-natural habitats of comparable conservation interest, but is more critical where semi-natural habitats occur mixed with improved grasslands. Generally the best approach is to audit the area as the majority habitat type, but describe the true complexity of the situation in the associated text. In fact this complexity might not be instantly apparent as the dimension of the various habitat patches may be seasonally dynamic. Patches of wetland may expand during the winter, and scrub and bracken encroachment and retreat is occasionally surprisingly rapid. The ecology of rapid plant community changes of this type is only poorly understood. The multiple auditing that is required to record this level of complexity (either remotely or on foot) is so expensive as to preclude it. Also it would be unlikely to add much to the quality of the resulting plan.

Auditing archaeological features

Most people carrying out farm environmental audits are unlikely to be expert archaeologists. For this reason many schemes require that this aspect of the audit is carried out by a regional archaeologist. They are typically overworked and the result is that this requirement can create a bottleneck in the process. Even if a particular scheme does not demand the involvement of a professional historian, it is recommended that they are consulted. Sometimes it will be sufficient to be given clearance that there are no aspects of archaeological significance on the farm. In contrast, farms with many historical features may require a joint visit with the local archaeological officer. When auditing a farm, even in cases with no recorded sites, it is important to look out for possible signs of archaeological remains. These signs include unusual lumps and ridges in the ground, or places where crops are shorter or slightly discoloured for no other obvious reason. If you are suspicious report the observation to the regional archaeologist. Producing management options for archaeological features is a job for a specialist. The approved prescriptions may sometimes involve measures that appear counter to those generally applicable. For example, scrub is usually protected by scheme rules, but it may require removing when its roots are considered as damaging to the belowground interests.

Auditing by remote sensing

Producing an acceptable farm environmental audit can be carried out by anyone following a basic amount of training. Clearly an agricultural background and some knowledge of natural history help, but the main habitat types and features that need auditing can soon be learnt. The process is, however, very expensive both in travel and by being labour intensive. Manual auditing by field officers raises concerns about repeatability between staff and can have problems of insufficient human resources being available at the right times of year. In addition, schemes officials may wish to audit farms throughout the duration of the management agreement to ensure that conservation work is being carried out and to monitor if the desired environmental changes are being accrued. To address all these concerns schemes are increasingly investigating the possibility of employing remote sensing to measure the amounts of on-farm habitats and monitor their environmental quality. The resolution of satellite images is now such that information can practically be extracted at the scale of approximately one square metre. Advances in information technology mean that data of this kind are easy and cheap to process. Wavelength analysis of the images can be used not only to identify different plant community types, but also to assess the amount of photosynthetic plant material present. This ability should make it possible to detect if hay crops have been taken before agreed dates and police other management options. This would involve repeat monitoring and this is currently still expensive. For now, many of these techniques still require ground testing, but they are certain to become increasingly important in future particularly as support for the agricultural industry moves from production and towards environmental protection. It is, however, difficult to imagine that remote sensing will ever be able to identify rare species or particularly fine examples of habitats from space. Furthermore, it is difficult to envisage farmers being enthused and encouraged to respect the conservation value of their holding by an aerial photograph in the same way that an experienced agri-environment officer can do on the ground.

Farm conservation plans and agreements

There can be a great deal of work involved in producing a full conservation plan, so before embarking on the process proper it is a good idea to produce a draft agreement with the farmer before investing time and effort. Your first priority must be to identify what the farmer wants from any particular scheme and how he/she sees the scheme working with his/her farming system. Motivations for entering schemes are diverse, but typically include preconceptions about what management options the farmer wishes to include in the agreement or capital projects, such as fencing or walling, he/she is hoping to fund. It is therefore important that you discuss these options with the farmer and decide if they are feasible options within the particular scheme.

One simple and practical way of producing a management agreement is to produce a 'shopping list' of all possible options available within the scheme. This may include alternative, mutually incompatible options for the same area of land. With each option, you can simply calculate the associated levels of payments and leave the farmer to select what he/she considers to be the most attractive set of options. This approach although simple may be unsuccessful in schemes where there is competition for funding between applications. Selection criteria are designed to maximise conservation enhancement and this approach does not consider this. Indeed, schemes differ in their entry requirements. Earlier schemes tended to be most generous, only requiring that the features detailed in the audit are not destroyed during the period of the agreement. Over time, scheme requirements have become more stringent and now frequently demand that a significant proportion of on-farm habitats or all habitats of conservation value are managed in an approved way. Whichever of these apply, you will need to discuss them with the farmer and hopefully arrive at a management plan that is acceptable within the scheme rules and agriculturally workable for the farmer.

Flood plains and common grazing

Many of the ecological impacts of agriculture occur at very large scales; these are discussed more fully in Chapter 9. This is likely to be particularly true in the uplands where reduced primary production means that top predators such as raptors will require large territorial ranges to ensure sufficient food. The lower input farming systems typically associated with the uplands are also likely to be associated with slower rates of environmental change. On top of these ecological aspects, land ownership in the uplands can also be more complex than in the more productive lowlands. For historical reasons, many upland areas across the globe are owned and grazed communally. These different elements can combine to make it extremely difficult to produce workable conservation plans in upland common grazings. Firstly, and most problematic, it is often difficult to reach a consensus agreement amongst graziers. There is no point in one farmer reducing stock numbers if another compensates by increasing them. This well-known phenomenon is termed the 'tragedy of the commons'. Although many agri-environment schemes contain upland options designed to incorporate common grazings, take-up levels have been generally disappointing because of the difficulties associated with trying to establish agreements between all the parties involved. This problem, coupled with the above ecological factors, means that large tracts of land and long time periods are needed before any agricultural or ecological benefit can be detected, and the outcome is hardly designed to attract participation in agri-environment schemes. These factors also in part explain why there is much less research published on the ecological impacts of agriculture in the uplands than the lowlands.

Similar issues of encouraging neighbouring farmers to participate in management options also apply to flood-plain and water-catchment management. Here again, desired environmental changes are only possible if all or most farms in an area are involved. Reinstating former flood plains can help prevent downstream flooding, as well as producing valuable wet-grassland habitats. Unfortunately, it is impossible for one farm to allow its fields adjacent to a river to flood without impacting on neighbouring farms. For this reason, when flood-plain management options are included within agri-environment schemes, they usually include restrictions (like those of common grazings), which insist that all farmers in the area must agree to the change and join the scheme.

In the case of water-catchment management it is not as critical that all farmers are involved. However, if the environmental aim is to reduce levels of nitrate within a watershed and river-system, clearly the more farmers involved in reducing their inputs and installing buffer strips, the better. In fact a number of agri-environmental problems exhibit these cumulative or threshold effects. Threshold effects occur when a significant environmental improvement can be achieved only after the conservation effort reaches a certain level. A nitrate load objective for a stream can only be met with the participation of a number of farmers within the watershed, or a level of habitat to sustain a target wildlife population can only be achieved with the participation of a group of farmers within a given area. To address these problems and to encourage coordinated involvement in target areas within agri-environment schemes, a number of projects have been established which operate within schemes. Typically, project officers from the local farming community have been given responsibility for liaising with other local farmers and informing them of the financial and environmental benefits of cooperative involvement in the scheme. The more successful examples of this have involved the support of local community leaders, or well-respected major landowners.

Conservation priorities in management planning

How do you decide what are the most important elements requiring conservation management on a farm? The answer to this will vary between areas, but there are some general guiding principles that are written into most schemes. Firstly, protect what is in good condition. Secondly, enhance what is in less than ideal condition. And only then think about habitat creation, hedge planting, pond digging, etc.

Existing features tend to be of greater conservation value than newly created ones because sustainable species diversity takes a long time to establish. The implication of this is that you must avoid the temptation to plant trees or dig ponds in old grasslands or wetlands. Much well-intentioned conservation work of this kind has resulted in damaging important habitats. You will frequently find that farmers have such projects in mind because they do not appreciate the conservation significance of existing habitats and see them merely as unproductive areas where a few trees could be planted. On the plus side they are generally easily dissuaded when they are better informed.

Conservation priorities change in different areas. It is important that you consider the wider landscape (see Chapter 9). What other habitats and species are found in the area? Do the on-farm habitats form part of a network of similar local habitats? Are there important sites directly adjacent to the farm? Conversely, does the site contain habitats or species that are locally rare? Many schemes formally recognise the importance of these elements and include them in the calculations that determine funding. Local conservation priorities may also be linked into schemes by local biodiversity action plans. There is no perfect or standardised methodology by which different schemes set the priorities that ration which plans are funded and which are not. However, there are a number of different approaches that can be taken.

Method 1. Points can be awarded for the number of different habitats managed

This method favours large farms over small ones, which can mean favouring intensive farms over less intensive ones (although this will vary with region and farm type). This system of prioritisation tends to encourage plans into introducing atypical elements, such as fields of arable production within livestock farms and vice versa, and thus favours more traditional mixed farming systems. Although this may be a good thing in conservation terms, it can be impractical for the farmers if they do not have easy access to the required machinery, and this might discourage them from entering into the scheme.

Method 2. Points can be awarded for the total area of habitats managed

This method again tends to favour large farms. It encourages the management of large blocks of habitat, rather than small isolated fragments, which is ecologically a sound thing. However, this approach may overlook the inclusion of important small fragments of habitat containing rare species in favour of large blocks of less pristine habitat.

Method 3. Points can be awarded for the proportion of land managed

In contrast to the above, this prioritisation system tends to favour small hobby farms, although the extent to which this is true depends on the scheme. In schemes where funding per farm is capped, large farms may be discriminated against because this limit may prevent a significant proportion of land being managed. Targeting small hobby farms may provide support for less productive land of higher conservation value, but perhaps does less in terms of achieving agri-environment scheme aims linked to social policies of supporting rural communities.

Method 4. Points can be awarded for including local priorities

Local conservation priorities may be identified by panels of experts, or following consultation with interested parties, or by using those identified in local biodiversity action plans. Including experts in the process may result in idiosyncratic priorities that may be difficult to practically include in a farm conservation plan. For example, local biodiversity action may identify rare species of hoverflies or lichens as being locally significant. However, it may be difficult to convince farmers of their relevance, and it is not always clear how plans can be modified to encourage these species. Conversely, consultation with interested parties tends to result in the same compromise local priorities applying over large areas, so that they are not 'local' priorities at all. A further drawback of this methodology applies when scheme funding is limited, and this results in priority inflation. When not all applications are funded, those applying know that in any particular year it is not worth applying unless you include X number of priorities (with the number being $X+1$ in the subsequent year), until all submissions incorporate all the priorities and the system fails.

Method 5. Prioritisation can be a hybrid of some or all of the above

As discussed above, no prioritisation system is perfect, but complex hybrid scoring systems may in part alleviate these problems. The downside is that hybrid systems can be very difficult to understand and administrate and may not be transparent if appeals are made. All systems of prioritisation result in applicants 'playing the game' of trying to amass the maximum number of points for minimal effort and this again can be problematic. Occasionally it may be that agri-environment scheme-approved management options are less than ideal for a particular sensitive habitat, and if the scheme is inflexible, then prioritisation points systems can encourage the management of sites that would be better left alone. For example, water-margin management options usually insist on stock exclusion, but water-margins that include species-rich grassland need grazing to maintain their conservation interest. Such problems are rare, either because schemes are flexible enough to avoid them, or because applicants can be discouraged from 'playing the game' to avoid damaging the specific conservation interest of their site.

What management options are available?

Generally the principle of 'manage the habitats that you already have', and as a lower priority create new habitats to replace what is likely to have existed before, answers this question. However, no scheme can have options available for all the infinite complexity of different farming systems that exist in the real world. Inevitably you will face questions about how to manage some unusual crop, for example are arable options such as beetle-banks applicable to this particular situation. These unusual cases will need to be dealt with on a case by case basis and will result in odd compromise decisions, such as swedes that are grown for human consumption being defined as an arable crop and thus eligible for beetle-banks etc., whereas swedes grown for fodder are considered forage and therefore arable options do not apply. Of course the environmental benefits would be identical, and this is another example of schemes being agriculturally rather than ecologically orientated.

When writing management prescriptions as part of a conservation plan it is important that you cover the minimal requirements of the scheme. Most schemes publish approved minimum management prescriptions. These detail, for example, maximum stocking rates or minimum stock exclusion periods. Management agreements can add extra restrictions when deemed necessary. These prescriptions are typically defined in agricultural terms such as by stock numbers rather than in ecological terms such as the condition of the vegetation. This may not always be ideal, as studies of hay-cutting dates have shown that annual variation in the onset of haymaking results in increased botanical diversity (Smith and Rushton, 1994). Unfortunately schemes encourage less variation in management, by enforcing standardised prescriptions, which are easy to police.

When writing prescriptions you should try to explain in language that is easy for the farmer to understand what management is required. For example, when setting acceptable stocking rates also quote them in equivalent stock numbers and types, also define the grazing periods and describe what the vegetation should look like before and after grazing. Furthermore it is a good idea to explain what the management is designed to achieve and how. The more the farmers understand the prescription, the more likely they are to apply it. Many farmers bend scheme rules, not out of malice or for financial gain, but out of ignorance of their significance.

Management plans

Most farm conservation management plans are agreed and documented in five-year blocks. Any shorter period and most environmental enhancements are unlikely to have occurred, any longer and it becomes impractical as a planning exercise. In fact, even the act of planning five years of agricultural activity ahead can sometimes be challenging. Most schemes require conservation plans that contain detailed description of the yearly conservation management and are linked to management maps. The management options are supported by annual payments related to the area or length of habitat being managed. To ensure that the maximum environmental gains are achieved, schemes usually insist that these prescriptions are initiated within the first year of the agreement. In addition to management options, schemes also include one-off capital payments. These are usually fixed-rate payments, either to cover the costs required to enable the management options to be carried out, e.g. fencing cost and alternative waterings when stock are excluded from streams, or they can be associated with one-off projects such as tree planting or pond digging. The first of these mean that there can be a large number of capital projects that need attention within the first few months of the agreement being implemented. In schemes that have a fixed date in the year when agreements start, this can mean it is very difficult to find a fencing contractor.

This factor may also result in large capital expenses for farmers with the associated payment cheque arriving some time later. For both of these reasons, it is probably best to spread the workload of capital projects not associated with management options over the full five years of the agreement.

In some traditional mixed farming systems, grass–clover forage crops are rotated with arable crops. During the rotation, soil fertility builds up because of the nitrogen-fixing ability of legumes within the grass ley, and this is released on ploughing to be exploited by a year or two of arable production. The final year of arable cropping may be under-sown with grass and clover and thus the cycle repeats. Rotations of this kind are generally regarded as having beneficial environmental impacts, providing valuable winter feed for farmland birds. They can, however, be very difficult to incorporate into five-year management plans, which need to define the exact amount of habitat to be managed each year, so that future payment levels can be calculated and budgeted for. Ideally the farmer would wish to monitor on a yearly basis the quality of grass–clover swards, and only at the end of the year decide which fields to plough and which can be kept for another year's grazing. However, if these fields are being entered into an agreement that includes a reduced input arable management option, then the exact area of arable land to be managed in this fashion in each of the next five years will need to be calculated before the start of the agreement. This might be good administrative practice, but it is not necessarily good agricultural practice. Furthermore, in farming systems of this type, where individual fields are unlikely to be kept in arable production for the full five years of an agreement, it might preclude them from management options such as beetle-banks, because schemes may demand that beetle-banks stay in place for the full five-year term. In cases where beetle-banks are allowed to move with the rotation, it can mean a lot of trouble and expense for the farmer in sowing them every year, and the management agreement will be complex because the areas being managed and the payments being received will change annually.

Rare and unusual species

The process of producing farm conservation plans is always varied; no two farms are ever identical in their conservation interest. However, the vast majority contain disappointingly little interesting habitat. Even in those that do, it usually comprises less than 5% of the total farm area. But every once in a while, you will discover something unusual; this will be more likely if you are an experienced field naturalist, but even to the untrained eye, interesting habitats likely to contain rare species are distinctive, even in the depths of winter. This should arouse your interest and cause you to look more carefully at the species present, and may require you to make a second visit to the site at a more appropriate season.

If you do stumble across an unusual or rare species, be it a rare orchid or butterfly, unless you are absolutely certain about its identification, you will need to have this authenticated by a local expert, preferably the local ecological recorder for that group of species. It may be a well-known site for that species or a new and exciting find. Contacting the local recorder is likely to arouse the concerns of the farmer. So before you do anything, speak with the farmer and reassure him/her. He/she is likely to have unfounded worries about masses of the general public swarming across his/her land to see/pick/photograph this rare species. It is important that you tell the farmer that the information will be kept confidential, and it is highly unlikely that anyone will visit the site, except perhaps an expert to confirm the identification. In reality most farmers are delighted and proud to be told their farm contains a rare species.

Next you will need to think about finding a management prescription compatible with maintaining this rare species. The good news is, however the site has been managed in the past, it cannot be far wrong. If vulnerable species have survived this long, whatever the farmer has been doing to the site must be compatible with the needs of the species. The chances are the current management will most probably be compatible with the scheme-approved management prescription. You may want to reassure yourself by discussing the management of the site with respect to that particular species with an expert. But the fact is, for most rare species, the optimal habitat management prescriptions are probably unknown. The expert is likely to ask you for details about how the site has been managed. In the unlikely situation where there is a discrepancy between the way the site has been managed in the past and the approved scheme prescription you may need to negotiate a special dispensation with the scheme administrator.

Site-specific issues

In addition to making your farm conservation plan compatible with a viable farming system and optimising the conservation benefits, there are many other site-specific issues that need to be considered. Many of these are associated with the planting of hedges and blocks of trees. For example, when drawing up a plan in the middle of summer it might seem like a nice idea to plant a hedge from the main road to the farmhouse. However, it might not appear such a clever idea in the depths of winter when the newly planted hedge acts as a barrier, dumping snow and blocking the road. With a little thought the planting of trees and hedges can be used constructively to prevent such problems, creating valuable shelter for livestock and helping to prevent soil erosion.

Other practical considerations

There are other easily overlooked issues that need consideration when locating trees and hedges that are associated with sporting interests. Many farmers interested in having conservation plans also have game shoots on their land. Poorly located blocks of trees or hedges can redirect flights of birds and ruin important drives. For this reason, the planning of tree planting on land where shooting is an important element is a specialist job, and one where it might be best to consult an expert. Similarly, tree planting along riverbanks can act as a buffer strip, prevent erosion, and help support invertebrates to feed fish, but in the wrong place it can disrupt fly-fishing activities. In some cases income from fishing can be considerable, and this may need to be balanced against agri-environment scheme payments.

Another common option available for conservation management that can bring about unexpected associated practical difficulties is the fencing of water-margins. Erecting stock fencing along a riverbank may be beneficial in terms of preventing erosion and hence improving water quality, but when the river is in flood, fences may trap debris and be washed away. A solution to this can be the seasonal use of electric fencing, which can be removed, along with the stock, in periods of heavy rain.

Agri-environment scheme agreement economics

The principle behind the funding of the majority of agri-environment schemes is that payments to farmers are carefully calculated to provide exact compensation for the reductions in income associated with lower agricultural productivity, rather than paying for the environmental gain achieved, although this principle may be eroded with time as interest in schemes outweighs the moneys available to fund them. For the most part, however, the net effect of this funding arrangement should therefore be that the farmer makes an economic decision to enter the scheme and in doing so sees no dramatic change in farm income. The reasoning behind this financial arrangement is that schemes are primarily and historically agricultural schemes, not environmental ones, and are associated with levels of agricultural production rather than environmental values. This has the advantage that agricultural production is easier and less controversial to measure than environmental value. Assigning an economic value to environmental goods and services, which are not bought and sold in a market, is difficult, expensive and at times controversial. Mechanisms were often in place to audit land's yield potential for earlier production-oriented schemes. This was not always the case, and some agri-environment schemes were designed, at least partially, as a method of providing financial support to poorer rural communities in more marginal areas. Some earlier schemes were relatively financially more attractive as a method of encouraging farmers into schemes when they were initially sceptical. Rates of uptake of early schemes varied widely, and often reflected different payment levels.

There are potentially serious environmental drawbacks to schemes that are linked to agricultural value rather than ecological value. For example, an existing species-rich grassland may contain many rare species in an ecosystem that has taken many years to develop. However, its productivity and agricultural values (and hence the payments its management might attract within a scheme) are low. In contrast, an area of former arable land sown as a new wild-flower meadow is likely to contain very few or no rare species. It will contain many common weedy species. It may be ecologically unlikely to persist, and may even act as a potential source of genetic pollution, contaminating local populations with alien genes. In spite of this low ecological value, such an area could attract large payments within an agri-environment scheme as compensation for the drop in income resulting from it being withdrawn from profitable arable production. The dangers here are obvious, and include the incentive to the farmer to convert important habitats into productive land prior to entry into schemes so as to attract higher levels of payments. Alternative systems, where payments are associated directly with ecological enhancement including paying specifically for the environmental goods and services provided rather than changes in agricultural management, or income forgone, are much more difficult to administer and monitor. Furthermore, they have the problem of how to financially reward a farmer who has diligently followed the scheme rules, but due to no fault of his/her own, this has resulted in no noticeable environmental change.

Using economics for the farmer's benefit

Fixing payments across nationally available schemes for particular management options, at face value, appears a fair and equitable system, and is relatively easy to administer and police. However, the true costs and real impacts of a particular change in agricultural management will in reality vary between regions and over time. A simple example of this can be illustrated by grassland management options designed to protect ground-nesting birds from being killed during silage making. Management prescriptions that offer payments to farmers for delaying cutting silage or hay until after the 15th May can be great news for hill farmers, or for those in the far north, who (except in the best of years) are unlikely to start haymaking before this date. However, for farmers in milder areas delaying the onset of haymaking until mid-May could be regarded as a disaster and they would require much more of a financial incentive to make this option attractive. Of course, not only is haymaking delayed in the upland, the nesting season is also likely to be later, so the environmental gains as well as the true costs of the prescription vary between farms. Some schemes recognise these sorts of differences, but it rapidly becomes complex if you start to insist on different cutting dates in different regions. In addition, these more site-specific schemes become extremely expensive to implement possibly resulting in less environmental benefit being provided for a given programme budget.

Not only does the value of agricultural production change between different areas, it can also vary significantly over time. This factor can be exploited by farmers within agri-environment schemes to maximise their economic returns, or alternatively if prices rise during the period of an agreement farmers might find themselves out of pocket. If annual management payments that are designed to compensate for reductions in arable crop yields are fixed during periods when grain prices are high, and are not adjusted within the scheme when grain prices fall, then such management options can become financially very attractive. Under these circumstances, applications have been known that include beetle-banks every ten metres across arable fields, because they are more profitable than the crops they replace. Another method of maximising the economic return gained from linear features such as beetle-banks and stream-side corridors results from payments that are paid for the nearest rounded up quarter hectare. Under these conditions, it always seems possible to find enough linear habitat to manage to ensure the payment always just falls into the next area category.

Not only do the economics of various management payments vary over space and time, they also vary between jobs. This simple fact can sometimes dramatically change the economics of participating in a scheme. For example, in most schemes capital payments are usually fixed at set levels irrespective of the actual costs involved. Therefore, if a capital project is simple and straightforward, for example erecting a long straight section of stock fencing, the work may be carried out by the farmer under budget and hence at a profit. However, if the same length of fencing is needed to cross uneven hilly ground, it will require lots of additional expensive corner posts, it may require the skills of an expert fencing contractor and the overall cost may well exceed the capital payment by a considerable amount. Similar variable costs can occur with most capital costs such as pond digging, and tree and hedge planting, and can result in conservation interests being compromised for economic ones.

Many agri-environmental schemes compensate participating farmers based on regional or even national average forgone income and/or capital costs. As discussed this approach is straightforward and inexpensive to administer. However, a significant disadvantage of this approach is that only those farmers with forgone income and capital costs less than or equal to average will be willing to participate in the scheme. This approach has been identified as cost targeting. With cost targeting, since the least expensive management agreements are provided the greatest number of agreements result. However, the pattern of adoption across the landscape may or may not provide the maximum level of environmental goods and services for the programme budget. Other targeting approaches such as benefit targeting, which targets agreements providing the greatest environmental benefits no matter what the cost, or benefit-cost targeting (greatest benefits at least cost) may be appropriate. However, these targeting approaches will generally be much more expensive to administer and will certainly provide fewer contracts or less area impacted than cost targeting. The targeting approach that provides the greatest environmental benefits for the given budget is likely to depend on the objectives of the programme.

Summary

This chapter has covered the practicalities of producing a farm conservation plan, and tries to integrate the ecological understanding covered in other chapters (1, 3, 6, 7 and 9) with workable agricultural systems. We have seen that schemes are likely to contain odd habitat definitions, they may incorporate overly complex prioritisation systems and sometimes the basic management prescriptions may not be perfect. This chapter has also discussed the problems associated with possible alternative approaches. It may be attractive to ecologists to produce schemes more tightly linked to formal plant community definitions and with management options and payment levels based on conservation value rather than stock numbers; however, with the limited human resources available to administer schemes and in the absence of agreed units of conservation value, then the current imperfect systems may be the best that is possible. As discussed in Chapter 3, research is starting to demonstrate slow but tangible increases in biodiversity resulting from farmers' participation in agri-environment schemes. So, although current schemes are by no means perfect they are starting to deliver some of their aims.

6

Habitat management

Introduction

The increasing use of inputs and the highly specialised types of farming systems practised nowadays have had a huge impact on both the agricultural landscape and species it supports (see Chapter 3). The abandonment of traditional farming systems and the expansion and intensification of agricultural production have resulted in the loss of many natural and semi-natural habitats. The integration of environmental objectives into agricultural policy (see Chapter 2) has encouraged the restoration of habitats of high conservation value and the creation of new ones. Various agri-environment schemes have been implemented where incentives are put in place to either encourage farmers to adopt management schemes that are environmentally beneficial, set-aside environmentally critical land or discourage or restrict management practices that are damaging to the environment (see Chapter 4). Enhancing the conservation value of agricultural land is currently attracting much research activity. This chapter reviews the scientific theory behind the maintenance of diversity, the assembling of communities and habitat management, and links between soil ecology and botanical diversity. Practical advice on what is possible/ acceptable on farms will be provided. The application of habitat management to alleviate the problems of fragmentation will be considered in Chapter 9.

What is conservation value?

In theory, if any form of land management is carried out for long enough on a particular block of land with a particular climate and geology then a predictable, recognisable community of plants and animals will develop.

In reality this is complicated somewhat by limitation to dispersal, the slow rate of ecological change compared with other environmental changes (such as climatic cycles or atmospheric nutrient deposition) and random disturbance events. Ecological communities that are associated with specific types of land management tend to contain predictable levels of diversity; some will support species that are currently rare or considered desirable/charismatic, while other communities will be made up of species that are common and thought of as weeds and pests. To an extent the 'conservation value/worth' that we ascribe to different communities and species associated with agricultural production systems is therefore a product of their current rarity and human preferences. Habitat types and species associated with native communities and now redundant agricultural practices are typically regarded as being of higher conservation value than those associated with modern, often industrial, practice. This generally reflects habitat rarity, and the ease with which a habitat can be recreated, but it also contains an element of historical artefact.

Diversity as estimated by species richness is a poor indicator of conservation value between different habitat types. For example habitats considered of conservation merit such as heathlands and moors may support lower levels of diversity than fallowed or set-aside arable land, which is generally considered of lower conservation value. It is a much easier task to compare the conservation value of two blocks of the same habitat type or to determine if one piece of land's conservation value is being degraded or enhanced over time, than it is to compare two different types of habitat. Within a habitat type species richness is just one of a range of recognised indicators that are used to prioritise conservation effort:

- Size
- Diversity
- Naturalness
- Rarity
- Fragility
- Typicalness
- Recorded history
- Position in ecological/geographic units/connectedness

Lists of indicators of this kind are widely used by conservation organisations to identify sites of high value (Ratcliffe, 1984) but such systems are much more easily defended within habitat types than between them. Sites are valued such that large sites are considered more valuable than smaller ones, diverse sites more so than less diverse ones, etc. Sites are also rated on a geographic scale in terms of their international, national, regional or local significance for species of interest. This element is often combined with the complex concept of rarity and used to assess conservation value at the species level. The strength of this approach is that it can ascribe high value to habitats and species at the edge of their ranges, which may contain important genetic diversity. However, it can also mean spending resources on conservation projects that would not be funded in other regions where the same habitat or species is widespread.

Within the agri-environmental context, habitats and species that are associated with agricultural practices that have become less common or abandoned by current managers have often declined in abundance with the intensification of agriculture and are generally seen as of high conservation status. Using this simple approach to assessing conservation value has resulted in species and habitats that were previously regarded as undesirable now being considered as being worthy of protection. What were once considered pest species are now actively encouraged; habitats such as wetlands that farmers previously received payments and advice on converting into 'improved' productive land are now the targets of habitat recreation grants. Under these circumstances it is not surprising that some question the meaning of the term conservation value. The logic of preventing the loss of species and habitats threatened by agricultural intensification is easy to defend. However, it is more difficult to rationally determine if there is an 'appropriate balance' which should be aimed at in the countryside, for example, between skylarks and magpies (using the first as an example of a species associated with more traditional British agriculture and the second as one that has thrived under intensive agricultural practices). Refining the definition of conservation value to this extent is problematic and meaningless unless constraints are added to the question, such as, what should the balance of skylarks and magpies be, given that the land is required to produce a quantity of food per unit area (as dictated by economic and rural development objectives) while also maintaining environmental function to meet environmental good and service objectives. Unfortunately we do not yet have adequate ecological understanding to answer such questions. Alternative methods of ascribing conservation value using economic measures including the estimation of the public's willingness to pay have been explored (Christie et al., 2006). Other approaches ascribe economic values to agricultural habitats based on the ecological functions and services that they provide to the human population, or relate to the cost of regaining the original environmental state or function following the loss of an area of habitat (e.g. Randall, 2002). Quantifying conservation value in monetary terms can be useful when evaluating alternative scenarios by representing the inherent trade-offs. However, valuing the environment can be very troublesome when there are many unknowns and uncertainty concerning the actual ecological function/true value of habitats and species, and is considered dangerous or unethical by many ecologists.

The principles of habitat restoration

Unlike habitat creation, which often has to tackle complex problems of altering soil chemistry and ecology to match some target community, in habitat restoration it is generally assumed the soil is already able to support a more diverse restored community. In principle the problems facing habitat restoration are twofold: firstly, there may be a need to reintroduce some species that have been lost or nearly lost from the site and secondly, there may be a need to reintroduce former agricultural management practices, or modify existing management practices, to fully reinstate the desired community.

The reintroduction of species itself raises two issues: what is the source of the reintroduced species and how do you introduce? The precautionary principle applies to the first of these in that it is considered desirable firstly to rely on natural reinvasion (establishment) of the site, and if this does not occur, to reintroduce local provenance material from a similar nearby habitat. This approach reduces the risks of introducing non-adapted genotypes, undesirable species and diseases, etc. as contaminants. With many species of plants the use of non-local genotypes has the potential to involve material with a different ploidy level to the local populations. Although unlikely, this possibility could have undesirable consequences and is best avoided by checking the source of donor species with adaptive ploidy level variation. However, generally the use of local provenance material must be balanced by higher costs and potential damage to the local donor site. However, there are management practices, such as spreading hay on the donor site, that can be adopted to reduce the damage to these sites. In principle the reintroduction of species as seeds (and the process nearly always involves seeds) is also straightforward. If seeds are sown over most pastures very few will successfully germinate and establish. To encourage seedling establishment it is usually necessary to create a 'regeneration niche', that is a gap within the sward large enough to reduce the competitive effects of the plants already present, but not so large as to expose the seedling to the rigours of the environment. In most cases such gaps are produced by the poaching/trampling effect of cattle or horses or by mechanical harrowing.

In practice most agriculture-based habitat restoration projects do not involve reintroductions, but tend to rely on reinstating traditional management as the sole method of diverting the community towards the desired composition. A good illustration of the need to reinstate all the different elements that were previously part of a traditional agricultural system in the process of habitat restoration can be seen in the work of Smith *et al.* (2000).

Table 6.1 illustrates that higher levels of botanical diversity (closer to the levels found in the desired target community) were associated with the

Table 6.1. *The values presented are the mean number of plant species recorded/4 m² after 4 years of hay meadow restoration in Britain after reinstating various aspects of the traditional management regime*

	Traditional management	Non-traditional management
No fertiliser added	17.9	
Fertiliser applied		16.4
Cut in June		17.8
Cut in July	18.2	
Cut in September		15.5
Grazed autumn & spring	19.6	
Grazed in spring		14.8
Grazed in autumn		17.1

Adapted from Smith *et al.* (2000).

reintroduction of each of the different elements of the community's traditional management. This habitat restoration experiment was established as a fully factorial design, and reintroducing all the different management options together was found to be most effective at increasing diversity.

It is generally the case that successful habitat restoration can be achieved in time solely by applying traditional management practices, although sometimes different tools are needed at least in the recovery phase, for example the use of higher stocking rates of cattle or horses to damage/open up the sward and facilitate seedling establishment. This can be an important component of early habitat restoration that would not be applied to a successfully restored habitat. Such interim management tools are often associated with reducing soil fertility and as such are more commonly used in habitat creation than habitat restoration.

While the restoration or creation of habitat that is dependent on some form of agricultural management can be accomplished while still meeting agricultural productivity objectives, the restoration of native habitat, such as natural wetlands, grassland communities or forest and shrub communities, often requires the explicit removal of management for agricultural production. Management prescriptions on these restored native habitats are developed to meet the specific environmental and ecological objectives of the site (Noss and Cooperrider, 1994). In many areas it has been necessary to remove land from agricultural production in set-aside programmes and, in some cases, establish ecological reserves in order to provide the quantity and/or quality of habitat required to sustain certain wildlife populations or to meet biodiversity conservation objectives.

Grazing as a tool in habitat restoration

Many agricultural habitats considered of conservation value are associ-
ciated with some level of grazing. However, the nature of the grazing employed,
in terms of types of livestock (species, breed, age and gender), when to graze and
for how long and at what stocking rate, all influence the resulting habitat
quality. As we have seen above, high-intensity grazing by heavy animals can
be important in producing regeneration niches early in the restoration process;
in contrast lower levels of grazing pressure are generally used to maintain
already diverse pastures. General guidelines on grazing levels for different
types of livestock have been produced for different types of pastures and different stock animals (Table 6.2). High grazing pressure is associated with rapid
removal of plant material, non-selective grazing and possible poaching/tram-
pling of the ground. This type of grazing can be useful in habitat restoration in
sites where invasive or unpalatable species have started to spread. In sites that
have been under-grazed for a year or two, high stock numbers can be used to
retard scrub encroachment; however, if this is too far advanced, cutting, burning or spraying may need to be employed. However, grazing can have direct
impacts on wildlife populations, for example high intensity grazing can result
in the trampling of eggs and nests of ground-nesting birds during the breeding
season. The probability of nest trampling varies with species and is directly
related to stocking rate. In contrast low-intensity grazing is associated with
slower rates of sward removal and hence slower structural changes in the
vegetation, which are beneficial for invertebrate populations, and ground-nesting birds are less likely to be trampled. Low grazing pressure allows selectivity
and tends to promote a more diverse sward by encouraging less competitive
unpalatable species. In certain range systems grazing management that
includes a period of rest for pastures (e.g. rotational grazing systems) has been
shown to provide certain plant species, particularly those that are highly pala-
table or those that are located in preferred areas (e.g. where water, forage and
cover are in close proximity), a period for recovery thereby helping to ensure
these species are not eliminated from the system by continuous grazing.

The development of agri-environment schemes has been associated with a
move away from traditional agricultural grazing practices based on assessment
of the sward and experience to systems based on recommended grazing rates
(based on standard livestock units) and defined periods of grazing. This approach
and the use of guidelines such as those in Table 6.2 are relatively easy to monitor
and justify to bureaucrats. However, there is a danger with such recommenda-
tions; grazing can quickly result in undesirable effects if weather conditions
change. The sward of pastures that become waterlogged can rapidly break up

Table 6.2. *The values represent the recommended numbers of grazing animals per hectare per time period in the United Kingdom needed to maintain good conservation value in different types of pastures. This can be achieved by high-intensity grazing over a short period or by lower stocking rates over long time periods. These different approaches are used at different stages in the habitat restoration process*

Number of grazing weeks	Calcareous pasture		Neutral pasture		Acid pasture		Wet pasture	
	Sheep	Cattle	Sheep	Cattle	Sheep	Cattle	Sheep	Cattle
2	60	15	100	25	50	12	–	12
4	30	8	50	12.5	25	6	–	6
10	12	3	20	5	10	2.5	–	2.5
20	6	1.5	10	2.5	5	1	–	1
36	3.5	1	5	1.5	3	1.5	–	–
Annual rate	2.5	0.5	4	1	2	0.4	–	–

Adapted from Nature Conservancy Council (1986).

even at low stocking rates, and conservation/agri-environment scheme-approved management options may need to be abandoned under such circumstances.

Grassland restoration

Grassland vegetation is dominated by plants from the *Gramineae* family. As a consequence of their growth form grasses are able to survive drought, fire and repeated defoliation by grazing herbivores. Grasslands occur naturally across the world, such as the North American prairies, where the growth of woodland is limited by low rainfall and high rates of evapotranspiration. However in many parts of the world, grasslands are semi-natural communities; the consequence of human activity. Following the last ice age, most areas of Europe were covered by woodland. During the Neolithic and Bronze Ages the systematic clearance of woodland by man began and under the influence of cultivation and grazing by domestic livestock grasslands became established. By medieval times grasslands were a dominant feature of the European landscape managed primarily for agricultural production. These grassland communities developed over a long period of time and their structure and floristic composition differ according to the ecological conditions. The interactions between climatic and edaphic factors (see Chapter 1 and Table 1.1) and biotic factors such as the type of grazing animal are very important in determining the type of grassland that occurs.

Over the last 50 years agricultural policy has encouraged the intensification of agriculture (see Chapter 2) and many aspects of modern grassland management. For example, the use of fertilisers, herbicides and pesticides, reseeding with new grass varieties, drainage and silage production have all resulted in the reduction of botanical diversity of existing grasslands and the wildlife it supports (see Chapter 3). A clear distinction can be made between those grasslands considered unimproved and managed under a traditional agricultural management and those grasslands that have been highly improved by recent agricultural management based on the species composition and richness of the sites. Modern grassland management techniques such as the application of fertilisers, ploughing and reseeding have all favoured a sward dominated by grass species and as a result most agriculturally improved grasslands are dominated by just one or two species. The management of these grasslands is designed to maintain the productivity of the grassland and the dominance of productive species. In contrast most unimproved grassland communities comprise a mixture of grass species and dicotyledonous herbs and are regarded as being of higher conservation value.

Enhancing the conservation value of agriculturally improved grassland is not always as straightforward as either idling the area or reinstating the traditional agricultural management as intensive grassland management could have considerably altered the grassland and soil system. The approach taken to grassland restoration will depend very much on the previous agricultural management and current environmental conditions as restoration will be affected by many factors including soil pH, soil moisture and nutrient supply along with the availability of seeds of the desired plant species (Berendse et al., 1992). The soil pH and soil moisture are instrumental factors in determining the type of grassland that can be restored. Identifying the most appropriate type of grassland for a particular site is an important first step in the restoration process. This is often complicated by previous agricultural management practices such as liming of acidic soils, field drainage and introduction of new and/or invasive plant species. Once the grassland type has been identified the most appropriate management for that grassland type can also be identified, for example whether the grassland should be cut for hay production or grazed by livestock, the type of livestock to be used, how many animals, when and for what period of time. However, the intensive application of nutrients may be a substantial obstacle to enhancing the conservation value of many agriculturally improved grasslands (see Figure 1.9 and the relationship between species diversity and soil fertility). In some cases the productivity of the site may have to be reduced before the traditional agricultural management can be reinstated. Frequent mowing and removal of hay has been used successfully to reduce productivity and promote grassland restoration (Bakker, 1989). In some cases greater intervention is required to reduce the fertility of the soil, this will be discussed in detail later on in the chapter.

Heathland restoration

The term heathland is used to describe vegetation that is dominated by evergreen dwarf-shrubs. Heathlands are found in various parts of the world: Europe, Canada, South Africa and South America where the ecological conditions (relatively cool temperatures, high atmospheric humidity and free-draining soils) favour the dominance of dwarf-shrub vegetation and trees are excluded through climatic, edaphic, biotic or anthropomorphic factors. In Europe, heathlands are found along the Atlantic coast of northern Spain, through the south-west of France and Brittany, the United Kingdom and along the west European coastal regions of Belgium, Holland, north-west Germany and Denmark, through southern Sweden to the Atlantic coast of Norway (Gimingham, 1972). Over much of Europe, heathlands have been derived from woodland and are essentially a semi-natural vegetation type owing their origin and existence to traditional land use and management, especially the grazing of cattle and sheep. Burning was used irregularly to prevent the spread of trees and to promote the growth of dwarf-shrubs such as heather (*Calluna vulgaris*). The burning of heathland for sporting purposes and the advent of grouse moors did not develop until much later, in the 1800s. In contrast, shrub-dominated landscapes in North America and Australia are created by arid conditions that create unfavourable conditions for grasses but can support deep-rooted shrubs (e.g. desert or xeric shrublands). The importance of heathlands to the farming system declined with the intensification of agriculture. Consequently the area of heathland has declined with much converted to productive arable land or pasture with the use of fertilisers. In many European countries the area that remains is greatly diminished. In Sweden an estimated 10 000 hectares remain compared with 150 000 in 1850. As most European heathlands owe their origin and continued existence to traditional forms of land use and management, they are potentially unstable and liable to quite rapid successional vegetation change when the management is changed or abandoned. The conservation of heathlands, therefore, requires active management. Changes in land-use patterns continue to threaten the existence of heathlands. However, other factors such as increasing nitrogen deposition (see Chapter 3) and lack of appropriate management are increasing threats. The cover of dwarf-shrubs in many heathlands is declining as increasing nutrient concentrations change the competitive balance, which results in the dominance of grass species. Overgrazing can also reduce the competitive ability of dwarf-shrubs, allowing a gradual increase in grass species, which are a natural component of the heathland community. In cases where management has been abandoned the increasing cover of scrub and woodland is resulting in a loss of heathland habitat.

The conservation of heathlands is important as they provide important habitat for a number of rare plant species, as well as birds, reptiles and invertebrates. Approaches to the restoration of heathland vegetation will depend upon the history of the site and whether any dwarf-shrubs remain. Where dwarf-shrubs are a minor component in the vegetation, the loss of dwarf-shrubs may be the result of frequent fires or overgrazing. Decreasing the grazing intensity may allow the surviving dwarf-shrubs to expand; however, if nutrient concentrations have increased this may favour the component grasses instead. Where no dwarf-shrubs are present but there is a substantial seed bank, restoration of heathland vegetation may be possible. Many heathland soils contain huge amounts of seeds of dwarf-shrub species, which can remain viable for many years. Such seed banks may be present under land converted to forestry. In most cases following productive agriculture no remnant vegetation or seed bank will be present. In this situation restoration is highly unlikely; however, it may be possible to create new heathland habitat depending on the intensity of the previous agricultural management.

Woodland and wood pasture restoration

The conversion of woodlands and forests to agricultural land and the use of non-native species for commercial timber production have increased the conservation importance of those areas of remaining woodland and forest. Within intensive agricultural landscapes woodlands often act as reserves for wildlife, but, unfortunately, many existing woodlands are unmanaged because it is not cost-effective as their products no longer have a market or in the case of certain non-timber forest products the markets are not large enough or adequately established. In many areas of the world, natural woodlands have been managed by humans for a long time for timber, grazing and other products. Consequently, one of the first questions in restoring woodland should be, does the woodland require active management? In Europe the answer is typically yes. However, it is perhaps naive to assume that all former woodland management practices were beneficial for wildlife, for example the removal of dead-wood for fuel, and the common grazing of coppiced woodlands may have reduced biodiversity. The objectives of woodland management will therefore vary depending on the age and type of the woodland. Woodland age is a good indicator of the

Table 6.3. *A selection of woodland vascular plants from southern England that are indicators of ancient woodland*

Latin name	Common English name
Allium ursinum	Wild garlic
Anemone nemorosa	Wood anemone
Campanula trachelium	Nettle-leaved bell-flower
Conopodium majus	Pignut
Hyacinthiodes non-scripta	Bluebell
Ilex aquifolium	Holly
Lamastrium galeobdolon	Yellow archangel
Oxalis acetosella	Wood sorrel
Paris quadrifolia	Herb-Paris
Potentilla sterilis	Barren strawberry
Primula vulgaris	Primrose
Prunus avium	Wild cherry
Ranunculus auricomus	Goldilocks buttercup
Stachys officinalis	Betony
Viburnum opulus	Guelder rose

conservation value and therefore the primary management aim of ancient or old woodlands will be conservation. A good method of assessing the age of a woodland or hedgerow is to look at the plant species within the woodland (Table 6.3). If there is a diverse mix of species within the shrub and ground layers of vegetation this is a good indicator that the woodland is old. Within these woodlands conservation management typically aims to maintain and enhance the existing species. The reintroduction of traditional management techniques such as coppicing can be beneficial. In newer woodlands, conservation may be a secondary aim to commercial objectives; here the restoration of a balanced age structure is important to ensure continuity of timber production. Woodlands and forests vary enormously in their species composition, age and previous management; consequently presenting detailed management prescriptions is beyond the scope of this book. There are many existing texts offering advice on how to manage woodlands.

Wood pastures differ from woodlands in that it is their structure rather than composition that defines what they are. It is estimated that forested land comprises 30% of the world's total area of rangeland. In many areas wood pastures are the result of historic management where large usually old trees occur within an open grazed landscape, which is typically grassland. They provide important dead-wood habitat for many invertebrates including species such as the stag beetle (*Lucanus cervus*). Many wood pastures have been lost through neglect and the loss of trees or conversion to intensive grassland or arable production. To maintain this important habitat and its associated species requires trees of different ages or a balanced age structure. However, many neglected wood pastures have few younger trees, often the result of overgrazing, to replace the mature trees in the future. In addition, the condition of many wood pastures has declined, as they have not been managed effectively as the mature trees have not been maintained properly. Techniques such as pollarding can increase the lifespan of a tree. In spite of this, pollarding is a rarely employed management technique nowadays. Restoration of wood pastures may require the planting of new trees to produce a balanced age structure and the reintroduction of tree management techniques to maintain the trees already present. The pollarding of very mature or ancient trees can be carried out although great care must be taken. For further information and advice on pollarding see Read (1996).

Riparian habitats

The majority of rivers within productive agricultural land will have been modified to a greater or lesser extent by humans. In many cases, the river channel will have been altered to take on the role of the flood-plain so that the flood-plain can be managed for agricultural production. Often the river channel will have been widened and deepened to accommodate greater flows during periods of flood. The consequence is often a simplified river channel, with steepened banks kept free of vegetation, which is too big for normal flows. Restoring riparian habitat is important for many factors such as bank stabilisation, the provision of habitat for birds and invertebrates, the input of woody debris and leaf litter, which can provide structure within the river, and shade, which can decrease the summer temperature of the water and help reduce excessive macrophyte growth. Agricultural management has often sacrificed riparian habitat for increased productive land by extending farming right up to the watercourse. Simply implementing a buffer strip by fencing off an area adjacent to the river can set in motion restoration of the riparian habitat, as a seed source of most plant species can usually be found upstream or *in situ* in the soil. If it is necessary to introduce seed, as with the other habitats discussed, native species and local provenance seed should be used where possible. The establishment of a mix of native riparian tree species, for example willow (*Salix* spp.), alder (*Alnus glutinosa*), hawthorn (*Crataegus monogyna*) and blackthorn (*Prunus spinosa*) in the United Kingdom, will help stabilise the river bank by root growth, and provide woody debris input (branches and leaf litter) and shade. If trees do not become established by natural colonisation then they can be introduced so they can provide these functions. There are many existing texts offering advice to riparian owners on how to manage their watercourse and how to maximise biodiversity whilst maintaining adequate drainage for agricultural production.

Nutrients are major sources of pollution in water as both nitrogen and phosphorus can be easily lost from farming systems (see Chapter 3). For example, the loss of phosphorus, mostly through soil erosion, is a particular problem in Europe as most freshwater bodies are phosphorus-limited. Excessive silt deposition is common in heavily managed rivers within agricultural landscapes. Intervention is often by dredging to clear this impediment to flood capacity. The spoils arising from this sometimes annual maintenance are spread on the bank top enriching the soil and can lead to vegetation dominated by strong colonisers; nettles (*Urtica dioica*), agricultural crops, oil-seed rape (*Brassica napus oleifera*). However, a major concern within riparian habitats is the rapid spread of non-native species. In the United Kingdom species such as Japanese knotweed (*Fallopia japonica*) and Himalayan balsam (*Impatiens glandulifera*) are problematic, while in Australia it is the spread of willow (*Salix* spp.), and in North America purple loosestrife (*Lythrum salicaria)* and saltcedar (*Tamarix ramosissima)* are two species that are causing problems. These species are classed as invasive species as they can become dominant very quickly and can easily overwhelm native plant species. In some situations where the availability of nutrients has been greatly increased by silt deposition it may be necessary to remove the topsoil from the riverbank to reduce the level of nutrients before riparian habitat can be restored. However, where topsoil has been removed there may not be an adequate seed source and species may need to be introduced.

Arable systems, beetle-banks, headlands, low inputs

Arable cropping systems are probably the most modified by humans of all agro-ecosystems, and therefore managing them for conservation, in many ways, is ecologically and technically the simplest. For the most part, managing arable land for conservation does not involve complex chemical adjustment of soil fertility. Arable production is usually associated with neutral pH and free-draining, fertile soil conditions, exactly the conditions that agriculturalists are experts in managing. Plus it is always easier to increase soil fertility than to reduce it. In addition, much of the arable ecosystem is assembled annually and does not require long periods of time for plant and animal communities to develop. If mistakes are made, the ecosystem can simply be ploughed up and a fresh start made in the next annual cycle. Of course this is overly simplistic and does not take into account complexities of soil seed banks or the time needed for populations of birds to respond to changes in arable farming. But the fact remains that arable systems are the most dynamic in terms of vegetation change, disturbance and nutrient fluxes of all agro-ecosystems.

In regions with long histories of agricultural production, arguably the largest negative impacts on biodiversity are associated with the intensification of arable farming and the conversion from mixed farming systems. In many of the economically and/or biophysically marginal regions, such as the wetter or hilly agricultural areas, arable agriculture has been reduced or removed, with modern crop production being concentrated into industrial scale intensive farm businesses. Thus, in these areas with long-term agriculture there are two very different aspects of managing arable farming for conservation that need consideration; firstly the reintroduction of arable cropping into former mixed farming enterprises and secondly reducing the intensity of production in specialist arable businesses.

The reintroduction of cropping into livestock farming regions is particularly important in producing winter-feeding habitats and spring nesting sites for many farmland birds. From an agricultural point of view it can help reduce feed and bedding costs if whole crop silage and straw can be produced on-farm. However, attempts to encourage such activity within agri-environment schemes may be limited because of a lack of relevant expertise by livestock farmers and because of the prohibitive costs of modern machinery used in arable production, which is designed for large-scale specialist units. This issue may be addressed by the use of contractors, or sharing machinery through cooperatives. Similarly issues of specialisation arise with modern crop varieties, which have been selected to be high yielding under optimal conditions and may fail to thrive in more marginal areas.

In extreme cases when arable production is being considered primarily as habitat creation for biodiversity then the cost and trouble of harvest can be avoided by feeding the un-harvested crop to livestock. Under these conditions, maximum conservation benefit can be gained by growing spring-sown cereals with low or no inputs. This produces a crop canopy, which provides spring cover for ground-nesting birds while not being too dense so that chicks are unable to dry out when it rains. Refraining from using herbicides and insecticides can decrease the negative impact on insect species thereby contributing to a more plentiful supply of invertebrates to feed these chicks. An extreme measure that can be taken in arable habitat creation for wildlife is to actively sow desirable and/or rare arable weed species. In contrast with the sowing of wild-flower seeds in grassland habitat creation, the cost is relatively low because 'weed' species are easier to cultivate and produce more seed. But unless the resulting plants successfully produce seed, this cost will be encountered on an annual basis. If colourful arable weed species such as poppies, cornflowers and marigolds are sown, the resulting crop can be very striking and such fields can be used as an advertisement for the more environmentally sensitive approach to farming being employed. The sowing of arable weeds as a form of habitat creation is only rarely used, but it is arguably one of the least management-intensive approaches and can generate quick results.

In regions where arable production is carried out on an industrial scale then the reintroduction of arable weeds is unlikely to be an attractive option. The speed and efficiency of farming activities in intensive arable production systems have had dramatic effects upon the species that live there. When vast areas are ploughed, sprayed or harvested in a few hours, there is simply no chance for wildlife to escape. Thus, slowing down or breaking up agricultural activities into smaller parcels could have beneficial effects for farmland biodiversity. Unfortunately the use of such mechanisms is financially unattractive. However, the strategic location and management of arable land taken out of production (set-aside) to reduce overproduction can provide refuges for wildlife and help increase populations of farmland birds (Bracken and Bolger, 2006). Within intensive arable areas the most widely used conservation management tools are beetle-banks and conservation headlands. These are newly created linear habitats, which are produced by sowing strips of tussock-forming grasses such as *Dactylis glomerata* (cocksfoot) and *Phleum pratense* (timothy). As with most grasses these species form more pronounced tussocks when sown at low rates. The ecological functions of these sown grass strips are to provide over-wintering habitats for beneficial invertebrate species such as rove beetles (*Staphylinidae* family) and money spiders (*Linyphiidae* family), which are predators of crop pests. The sowing of grass strips around field-margins can help suppress arable weeds and prevent fertiliser and pesticide drift. Preventing spray drift into field-margins reduces the extent of disturbed fertile habitat available to support undesirable weed species. Since field-margins are the lowest yielding areas within arable fields, surrendering such land to beetle-banks or conservation headlands can often have limited impact on overall profitability. However, since beetle-banks (simple sown strips of tussock-forming grasses) can also be located across the most productive central parts of fields, the economics of habitat creation here is less attractive. Not surprisingly, therefore, beetle-banks are much more commonly found in field-margins than across arable fields. Conservation headlands can be considered as a refinement to the simple beetle-bank habitat (Figure 6.1). In addition to a sown grass-margin, conservation headlands also incorporate an unsprayed or low-input headland strip of cropped land. These have been specifically designed to encourage game bird species such as grey partridge (*Perdix perdix*). Chick survival rate is increased, as more invertebrate food is available

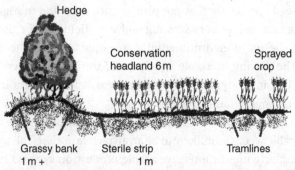

Hedge

Conservation
headland 6 m

Sprayed
crop

Grassy bank
1 m +

Sterile strip
1 m

Tramlines

Figure 6.1 A conservation headland, consisting of a sown grassy bank as habitat for invertebrate generalist predators and to suppress weeds, a sterile strip to hinder weed ingress into the crop and a low-input cropped headland designed to increase the survival rate of game bird chicks.

within the headland. Again, due to the fact that these management schemes take land out of production, in most cases they are economically viable only on the most marginal land. However, encouraging weed growth even in the headland of a crop can have deleterious effects on grain purity and moisture content at harvest. These problems can be addressed by late applications of a general herbicide such as glyphosate after the crop is mature and the habitat has served its purpose, although this approach is not feasible with malting barley as it kills the grain. An alternative that can be used in this case is to harvest the headland separately and use the grain as feed for stock or rearing game birds.

The discussion above focuses on management schemes that can enhance habitat on arable land in regions and countries (e.g. Europe) where agricultural production has a long history, and as a result the extant wildlife species are adapted to intensive agricultural ecosystems. However, management prescriptions to address habitat requirements on landscapes with a shorter history of agriculture (e.g. North America, Australia), where many of the extant wildlife species require native plant communities to meet at least some of their habitat needs, may involve some different approaches. For example, many species of ducks and shorebirds in North America are dependent on natural wetlands as well as areas of native uplands that remain within the arable agricultural landscape. Therefore, preservation of these habitat types in many cases requires a shift from management for agricultural production to less intensive production systems, land idling or specific management as wildlife habitat. For example, in western Canada, wildlife habitat initiatives often focus on setting aside priority habitat parcels and decreasing the intensity of agricultural production in areas adjacent to these habitat areas (e.g. riparian zones and other wildlife corridors). The objective of many of these schemes is to provide habitat characteristics that are similar to pre-agricultural landscapes.

Potential obstacles to habitat creation on farmland

The creation of semi-natural habitats can only be achieved if the appropriate ecological conditions for that habitat are present together with a supply of propagules of the target species. Agricultural land is not always the ideal environment for the creation of semi-natural vegetation as agricultural use changes the edaphic conditions and the availability of the target species within the agricultural landscape is usually limited.

Nutrients and soil fertility

The intensification of agricultural practices and the increased use of fertilisers over the past 60 years (see Chapter 3) have increased the concentrations of nutrients in farmland soils. When comparing soil from farmland and semi-natural habitats the concentration of available nutrients such as nitrogen and phosphorus is much greater in the agricultural soils (Gough and Marrs, 1990; Pywell et al., 1994). The high concentrations of nutrients found in agricultural soils are recognised as a major obstacle to habitat creation on farmland as most semi-natural habitats occur on soils of low fertility. The addition of nutrients to semi-natural habitats can change the floristic composition, above-ground biomass and species richness of the vegetation (see Figure 1.9 and the relationship between species diversity and soil fertility). To establish semi-natural communities on farmland it is often necessary to reduce the concentration of nutrients in the soil, for which there are several techniques available (Marrs, 1993). Direct removal of the nutrient pool using methods such as topsoil removal has been very successful in decreasing the soil fertility as the concentration of nutrients is often greatest in the top layer of soil. Other suggested methods include removing nutrients through the crop (continuous cropping) or by exporting hay. Nonetheless, targeting habitat creation to those agricultural sites with a low nutrient status already is likely to be the most successful approach. In situations where this is not feasible due to other considerations, such as extending existing habitats and reducing fragmentation (see Chapter 9), management to reduce the concentration of nutrients may be necessary.

To create semi-natural habitats on farmland requires propagules of the original plant community to be available. The two important sources of propagules are the seed bank within the soil or by dispersal from neighbouring habitats (seed rain). The seeds of different plant species vary in their ability to survive in the soil. Plants of disturbed habitats usually have seeds that are relatively long-lived in the soil while plants of undisturbed habitats form seed banks that are transient in nature with few surviving from one year to the next. Farmland soils usually have a soil seed bank dominated by annual and ruderal species, which are able to survive for long periods in the soil. The establishment of semi-natural habitats from the soil seed bank is highly unlikely after intensive agricultural management as the required species are no longer present (in sufficient numbers) in the soil and any annual and ruderal species that are present are potential competitors. Plant species associated with semi-natural habitats are usually poor dispersers. The dispersal of seeds is limited to only a short distance from the parent plant; therefore, establishment of new habitats is usually limited to sites neighbouring existing sites where there is a source of propagules. However in most situations leaving a site to colonise naturally is not feasible and the introduction of species is necessary. The introduction of species to a site can be done using a variety of techniques. Sowing of seeds that are harvested from existing sites or supplied by nurseries is the most commonly used method, although the application of hay or litter and the transfer of turf or soil from donor sites are being used more frequently. Each method has it own advantages and disadvantages (Edwards et al., 2007). The harvesting of seeds or hay from local sites overcomes the issue of local provenance. The use of seeds allows for the removal of unwanted species such as thistles or docks by the process of seed cleaning. Seeds can also be stored, while hay and turf need to be used immediately.

Grassland recreation

The issue of local provenance has been referred to earlier in the chapter; however, it is a particular issue in grassland creation as many grassland species are known to show geographic genetic variation. The flower colour polymorphism in *Lotus corniculatus* is a good example. Populations in the north and east of the United Kingdom have a greater proportion of plants with dark keels, whereas the yellow keel prevails in populations from the south and west (Crawford and Jones, 1988). Within the grassland context genetic provenance is known to relate to management as well as local climatic conditions, with flowering times being earlier and more tightly constrained in hay meadows than in grazed pastures. Grassland communities that extend over relatively wide latitudinal or longitudinal ranges can be made up of subpopulations that are adapted to widely differing climatic conditions (e.g. precipitation, heat units). However, for many plant species the detail and distribution of geographical and adaptive ecological variation is not well known. It is possible that introducing plants from other sources will adversely affect local populations. Firstly, local plant populations may be better suited to the local conditions or secondly the introduced population maybe superior (Bischoff *et al.*, 2006). Where feasible, preference should be given to the establishment of grassland species by natural colonisation, but in situations where this is not possible local seeds should be used, taken from the same type of habitat (Hopkins, 1989; Bischoff *et al.*, 2006).

The introduction of plant species by sowing raises some further issues. How many species should be sown? And which species should be sown? One of the main aims of grassland creation is to recreate species-rich grasslands that have been lost through the intensification of agricultural practices. Sowing a species-rich mixture can be very expensive and for that reason is it always necessary to sow a mixture with a large number of species? In most cases it does appear that sowing a seed mix with a greater number of species will be more successful as the initial seed introduction is a very important factor in determining the composition of the grassland (Egler, 1954). Introducing more species also acts as insurance against failure (Yachi and Loreau, 1999). The choice of species to be sown is usually determined by the type of grassland to be created, which is usually influenced by environmental conditions such as soil pH. However it appears that particular plant traits are also important in determining the success of restoration (Pywell *et al.*, 2003).

Although critical to the success of grassland creation, the availability of seed is not the only factor that determines the successful creation of species-rich grassland. Failure to germinate and become established, from either an introduced or natural source, as a result of unsuitable environmental conditions will lead to failure. It is therefore necessary to understand the ecological factors which prevent successful establishment. As mentioned earlier in the chapter, agricultural land is not always the ideal environment for the creation of semi-natural vegetation as agricultural use increases the nutrient concentrations in the soil. Choosing sites where the residual soil fertility is comparatively low to reduce the effects of competition is more likely to result in success. In addition, farmland soils usually have a soil seed bank dominated by annual and ruderal species. Weed control prior to the introduction of grassland species has been shown to be very important to the success of establishment (Lawson *et al.*, 2004a). While the problem of high soil nutrient concentration in agricultural soils has been acknowledged for some time (Marrs, 1993), recent research suggests that biotic properties of the soil, such as the soil microbial community and soil fauna, may be equally important in influencing community development (Smith *et al.*, 2003; De Deyn *et al.*, 2004). For example, land that has been used to produce annual crops will over time alter the soil biodiversity, often losing the soil fauna that is required for the growth of the native species.

Moreover, it has been shown that individual grassland plant species can affect the establishment and productivity of other grassland plant species through changes in soil properties (Bezemer *et al.*, 2006). Understanding plant–soil interactions and how these processes affect community assembly are key to improving the success of grassland creation.

Heathland recreation

Heathlands are typically found on nutrient-poor acidic soils, which have a pH ranging from 3.4 to 6.5 (Gimingham, 1972). As a result two key factors associated with farmland have been identified as potential problems for the creation of heathland vegetation. The addition of fertilisers and lime, which significantly changes the soil conditions; raising the pH and elevating the nutrient status and ploughing, which changes the structure of the soil; mixing the well-defined mineral and organic soil horizons, which destroys the typical heathland soil profile (Pywell *et al.*, 1994). To create the appropriate edaphic conditions for the creation of heathland on agricultural land steps to reduce the pH of the soil and concentrations of nutrients are usually necessary. Various methods such as topsoil removal, cropping treatments and the addition of sulphur have been used to reduce the levels of nutrients and soil pH on agricultural land to aid the establishment of heathland species (Marrs, 1985; Marrs *et al.*, 1998; Owen and Marrs, 2000; Lawson *et al.*, 2004b). In most cases the seeds of many heathland species are unlikely to be available from either the soil seed bank after intensive agricultural production (Pywell *et al.*, 1997) or by dispersal. For that reason it is often necessary to provide a source of heathland propagules. Heathland species can be introduced to a site using a variety of different methods: sowing seeds, applying harvested shoots or litter (the plant material overlying the soil) or transferring turfs. Successful heathland creation depends very much on establishing the appropriate ecological conditions without which heathland species will either fail to establish or persist under competition from other species. Ideally heathland creation should be targeted to sites with an acidic soil, where the concentrations of nutrients are low and where heathland species are available. Sites where there has been a past history of heathland or which are adjacent to or link existing areas of heathland are likely to achieve the maximum conservation benefit.

Woodland recreation

The first consideration in woodland creation is the choice of site for the new woodland. Care must be taken to avoid areas that already have a habitat with high conservation value such as heathland or species-rich grassland. Sites

Table 6.4. *Main types of beech woodland found in the southern United Kingdom determined by soil conditions*

Woodland type	Soil types
Beech–ash woodland with dog's mercury	Calcareous
Beech–oak woodland with bramble	Mesotrophic
Beech–oak woodland with wavy hair-grass	Acidic

Adapted from Rodwell (1991a).

adjacent to existing woodlands are likely to be the most successful creation sites as they can provide a source of plants and animals to colonise the new woodland. Natural colonisation is preferable to the planting of trees and shrubs; however, like the other habitats discussed in this chapter, natural colonisation is not usually the most suitable option for woodland creation. Once a site has been chosen, the next step is to identify the most appropriate type of woodland for a particular site and the most suitable species of trees and shrubs to be planted. Climate and soil type are the main factors determining the distribution of woodland types. For example in the United Kingdom, beech (*Fagus sylvatica*) woodlands naturally occur in the south on free-draining mineral soils, and where the soil type determines the distribution of the accompanying species (Table 6.4). Where possible trees should be planted within their natural distribution and the soil type should be determined before the choice of species is made to ensure the created woodland closely resembles the native woodland type. However, this is often complicated by the previous agricultural management as the liming of acidic soils and field drainage may have taken place.

In creating new woodlands it must be remembered that woodlands have several layers of vegetation. Introducing shrubs, as well as trees, and encouraging the development of the ground flora will ensure the species composition and structure of the new woodland is similar to that of existing woodlands. Sites that already have some species of the ground flora present are more likely to succeed. In creating new woodlands several other factors should also be considered such as the density of planting and the mix of different species. For further reading on creating new woodlands see Rodwell and Patterson (1994).

In many regions there are increasing areas of agricultural land being allocated to agroforestry management, while agroforestry can represent a broad range of economic objectives involving crops of trees. In many cases these tree stands are single species stands (e.g. poplar (*Populus* spp.), eucalyptus (*Eucalyptus* spp.)) that will be harvested for a timber product such as lumber or pulp and paper. Some of these stands are also being established as carbon sinks to assist countries in meeting their greenhouse-gas emission reduction commitments under the Kyoto protocol (see Chapter 10). These forest stands may provide only very limited habitat benefits as the management may be very intensive, including the removal of all competing species, and maintaining straight rows of trees with open lanes to access the trees for ease of management and harvest.

Summary

In landscapes dominated by agricultural production, agricultural management may play a very important role in maintaining and enhancing biodiversity. Increasing the area of agricultural land under management to meet environmental objectives, along with the restoration or creation of habitats of high conservation value, can make an enormous contribution to achieving biodiversity objectives, by increasing the size of remaining habitats, linking fragmented areas together and creating areas of intrinsic value. The long-term aim of restoration or creation is often the establishment of a particular habitat, such as species-rich grassland. The major obstacles to the restoration/creation of habitats of high conservation value on agricultural land have been well documented. Short-term intervention, for example sowing seeds of appropriate species and depletion of soil nutrients, is often considered necessary to achieve the long-term aim. Often management is concerned with the vegetation and little consideration has been paid to the associated species. It must be remembered that any management undertaken should also create favourable conditions for the species associated with that habitat. In some situations the best management practice may differ between the vegetation and associated species (Woodcock et al., 2006) or differ between groups of species. Where this is the case management will depend on the conservation objectives of the individual site.

7

The management of agricultural wastes

Introduction

In many respects agriculture can be regarded as any other industry, in that it uses resources to produce products and in doing so it produces by-products that are potentially polluting wastes. However, unlike other industries, the products of agriculture can be divided into foods and materials, which are directly purchased by the consumer, and custodianship of the land, which historically has been unappreciated and not recompensed. In the later part of the twentieth century, the role of farmers as custodians of the countryside was first fully appreciated and this was typically manifested in two ways. Firstly, agri-environment schemes were introduced as a mechanism by which farmers could receive payment from the state to actively manage their land in ways considered to be beneficial to the environment (see Chapters 4 and 5). Secondly, legislation was increasingly introduced which regulates the disposal of agricultural waste products and hence protects the environment. Even in the absence of legislation, agriculturalists have a moral obligation to dispose of their wastes responsibly. Pollution incidents can only result in the image of the industry being damaged in the eyes of its increasingly aware consumers. This chapter therefore reviews the range of waste materials that are produced by modern agriculture, the potential environmental impacts that can result if they are not dealt with appropriately and options available for waste management planning for efficient resource use and meeting current and future legal liabilities.

What are farm wastes?

The list of materials that are produced as by-products of agricultural activity is long and diverse and contains many substances that have the

Table 7.1. *Classification of farm waste products*

	Solid wastes	Liquid wastes	Gaseous wastes
Materials produced on-farm (tend to be biological)	Crop residues Fallen stock (dead livestock) Manures Spoilt feedstuffs	Slurry Silage effluent Spoilt milk Vegetable washings	Methane Ammonia Nitrous oxide
Imported materials (tend to be inorganic materials)	Packaging, pesticide containers and silage wraps Scrap machinery, batteries, tyres, etc. Waste building and fencing materials	Oils Pesticides and spent sheep-dips Waste veterinary products	Carbon dioxide

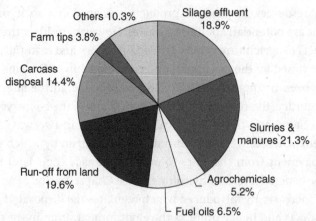

Figure 7.1 Breakdown of farm pollution events in Scotland. (Adapted from SEPA, 2000)

potential to be very damaging to the environment. Within most legislative systems, the full range of farm waste products are poorly defined, with specific laws applying to specific materials. However, farm wastes can be classified into five general categories (see Table 7.1).

Of the farm waste types identified in Table 7.1 those produced on-farm have received most attention in terms of legislative regulation, research activity and the development of specialised management systems. This relates to the fact that they are produced in larger quantities and typically have greater polluting potential than imported wastes (see Figure 7.1). For this reason this chapter will primarily focus on this group of waste materials.

Farm-produced wastes/nutrient management

For the most part the management of those wastes produced on-farm is concerned with regulating the movement of nutrients within the agro-ecosystem and for this reason it should not be referred to as waste management at all but more correctly termed nutrient management. Within agro-ecosystems nutrients are much more dynamic than within natural ecosystems. On-farm there are three main reservoirs for nutrients, plants, livestock and the soil, with stored manures and slurries arguably representing a fourth. It is relatively easy for nutrient cycles to become distorted within agro-ecosystems because the nature of farming is to export nutrients out of the system in the form of food for human consumption. In most modern agri-cultural systems it is rare or even illegal for nutrients to be returned from the human food chain back to agricultural land and this loss is typically compensated for by the purchase of artificial fertilisers and concentrate feeds. Balancing nutrient levels is complicated by a host of factors including: crops having different requirements (both temporal and absolute), which vary with the nature of the growing season, soil type, gradient and water avail-ability, all of which vary between nutrients. Nitrogen has generally received most attention because it is the most mobile of the nutrients and hence is most difficult to manage.

The literature contains many variants of nutrient cycles differing in complex-ity. However, they all purvey a similar view of an ecosystem function that can be traced back to Aristotle. The implied underlying principle is that there are mechanisms at work in the world (originally considered to be derived from god) that ensure order and balance (Figure 7.2). Such views of natural balance are strongly held within the organic movement. However, there is nothing implicit within the structure of a nutrient cycle that means that equilibrium states will be maintained. On a global scale only in the late twentieth century did the amount of atmospheric nitrogen fixed by human industrial activity as fertiliser exceed that fixed by the action of lightning. Currently worldwide approximately five times as much nitrogen is fixed per year by the Haber–Bosch process for agricultural use than is fixed by lightning (Figure 7.3). This massive perturbation in the nitrogen cycle is considered by many ecologists as large-scale nitrogen pollution of the environment. It has also been argued that this sudden growth in global fertiliser consumption has driven the growth in the human population (Smil, 1997). Even the increase in organic fixation of nitro-gen, resulting from humans cultivating leguminous crops, can be considered as a dangerous change in the dynamics of the system which has the potential to have considerable ecological implications.

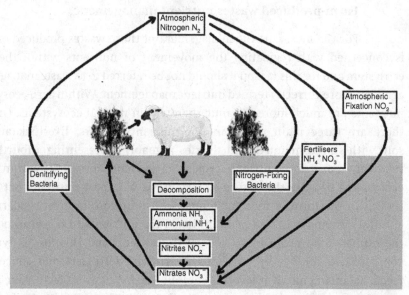

Figure 7.2 This apparently completely closed version of the nitrogen cycle omits losses to the agricultural system via the export of products or via run-off and erosion.

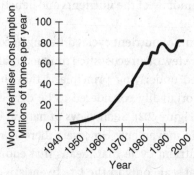

Figure 7.3 The production of 'artificial' nitrogen by the Haber–Bosch process massively distorted the nitrogen cycle during the second half of the twentieth century. Many habitats can now be considered as being polluted with nutrients from agricultural activity.

In intensive purely arable farming systems, typically 90% of nutrient inputs are in the form of artificial fertilisers. This results in the nutrient cycling process being much simplified, while flux rates through the system are increased. Depending on location, between 58% and 87% of the available nitrogen and phosphorus may be removed during harvesting and often considerably more potassium is exported in the crop than is replaced as fertiliser (PPI, 2002). With high artificial fertiliser input rates, there is little need for bacterial or fungal decomposers within the soil in intensive systems. Since the release of nitrogen by decomposer microbes can be limited by temperature and application of fertilisers is not, this change means that crop growth may start earlier in the season. Under these conditions the ratio of bacteria to fungi within the soil is shifted in favour of bacteria, as complex mycorrhizal symbiotic relationships between plants and fungi no longer function. The soil ecosystems of intensive arable systems support fewer invertebrates as soil organic status declines over time. Even so, ploughing can still stimulate bacterial activity resulting in a pulse of nitrogen release producing the potential for leaching. With such high nutrient fluxes occurring there is a high risk of run-off, and therefore central to successfully managing the use of high fertiliser inputs is ensuring that ploughing, application rates and times match the requirements of crop growth. Even when fertiliser applications are synchronised with crop growth, recommended practice is to avoid spreading too close to watercourses and to establish buffer strips of vegetation designed to intercept leaching nutrients. In areas considered to be at risk of nitrogen pollution (Nitrogen Sensitive Areas) recommended practice might even include the shattering of land drains adjacent to streams and rivers to allow the interception of nutrients that would otherwise be unavailable to buffer strip vegetation. The effective management of such Nitrogen Sensitive Areas requires considering nutrient flows at large landscape scales by integrating the activities of several farmers (see Chapter 9). A dramatic example of the impact of landscape scale nitrogen run-off is the hypoxic zone in the Gulf of Mexico, which attained a maximum measured size of 22 000 km^2 in 2002 (USGS website). In this hypoxic zone water contains less than 2 parts per million of dissolved oxygen causing fish to leave and stress or death to bottom-dwelling organisms. This hypoxic zone is caused primarily by excess nitrogen delivered by the Mississippi River primarily from run-off from agricultural activities in the central United States.

Historically levels of nutrient inputs have been lower in grazed pasture systems than in arable. Indeed, traditionally animal manures were used in mixed farming systems to move fertility from more marginal land to the better fields used for arable cropping. During the twentieth century the increased reliance on improved pastures, associated with increased levels of artificial fertilisers and the increase in use of concentrate feeds, resulted in a reversal in the movement of nutrients from arable land back to grazed pastures. Grazed pastures are associated with higher levels of nutrient cycling than are non-grazed swards because of the greater efficiency of nutrient transfer from the vegetation to the soil that occurs via faeces and urine and because of the lack of accumulation of organic material on the surface. Herbivores use a relatively small proportion of the nutrients that they ingest, and typically return 60% to

99% of the ingested nutrients to the pasture. The majority of biological cycling of nitrogen that occurs in grazed pastures does so via urine rather than faeces. Inputs of labile nitrogen as urine are equivalent to 300 to 600 kg N ha^{-1} yr^{-1}, with 30% to 40% of the pasture surface receiving a urine 'application' during the year. In contrast, the nitrogen levels excreted through the faeces of sheep and cattle are usually about 0.8 g N 100 g^{-1} of dry matter consumed regardless of the nitrogen content of the feed (Haynes and Williams, 1993). Although livestock may return up to 99% of the nitrogen they consume to the pasture, as much as 60% of this can be lost via volatilisation to the atmosphere, whereas approximately 90% of the consumed phosphorus and potassium are returned to the pasture and are retained in the soil. However, in grasslands that are harvested as hay or silage, this return of phosphorus and potassium does not occur and there is therefore an increased potential for nutrient imbalance to occur. The permanent cover of vegetation in pastures means that the leaching of nutrients and soil erosion are less of a problem. However, if manures or slurries are applied to pastures at inappropriate times, rates or weather conditions there is still great potential for the loss of valuable nutrients and considerable ecological damage. Examples of this are not infrequent and can be seen in Figures 7.4 and 7.5. Such incidents typically occur after periods of heavy rain when waste storage facilities are inundated with water. This can be avoided if grey water is separated from slurry storage, and surfaces contaminated with manures are kept to a minimum and under cover.

In many parts of the world, mixed arable and livestock farming has been in decline. In such mixed farming systems, including most organic farms, nutrient cycling tends to be more closed and hence more manageable. Central to such farming systems are periods of fertility building using nitrogen-fixing crops integrated within a rotation and lower levels of nutrient inputs from outside the farm. The use of nitrogen-fixing legumes rather than fertilisers results in less dramatic fluxes of nutrients and typically lower levels being involved. However, the potential for leaching remains if cultivations are mistimed or when crop growth is delayed by poor weather conditions. Therefore, the selection of an appropriate rotation can be critical to reducing the pollution potential of agriculture. Concerns have been raised by some conservationists that the nitrogen status of low fertility species-rich grasslands may be compromised by organic farmers slot seeding with clover in the misguided view that naturally fixed nitrogen is ecologically benign.

Extensive farming systems associated with more marginal land (e.g. low fertility soils, more arid areas, hill land and highly erodible soils) are now often based on the grazing of rough grasslands or dwarf-shrub communities, although historically some arable production was an important component. The

Figure 7.4 This photograph shows an example of poor management of slurry, where liquid waste was simply pumped into a pasture. The result was a loss of grazing and nutrients, a subsequent increase in undesirable weed species and a decline in botanical and invertebrate diversity. In the short term, however, the site was used extensively by feeding snipe.

amount of nutrients in circulation under these conditions is low in comparison to other types of farming and nutrient release by decomposing microbes may be restricted seasonally by lower temperatures. The potential for nutrient pollution via leaching is exacerbated by elevated levels of precipitation in the hills but this is compensated for by lower stocking rates, which reduce the amount of faeces and urine deposited per unit area. In addition low pH, waterlogging and low levels of invertebrate activity in upland soils all contribute to reduced decomposition rates and a consequential build-up of organic material in the soil.

Farm waste management planning and nutrient budgeting

In many livestock production systems in temperate regions, animals are housed for a proportion of the year. The consequence of this is that animal manures and slurries are produced that are not directly deposited by the animal to land, so that the farmer is responsible for collecting, storing and utilising these potentially highly polluting materials. In most countries there are codes of good practice that cover the handling of animal wastes and in many these are

Figure 7.5 This photograph shows another example of poor slurry management, where liquid waste spread was over waterlogged ground. This resulted in damaging the soil structure and leaching of most of the nutrients involved.

reinforced by legal constraints. In addition to handling animal manures and slurries, waste management planning must also account for the disposal of other organic wastes such as waste milk, silage effluents and vegetable washings. All of these are potentially highly polluting (see Table 7.2) and are usually disposed of by mixing with slurry. However, silage effluents can provide a valuable carbohydrate source for livestock, if they are kept separate and uncontaminated from other wastes. Alternatively the volumes of silage effluent produced can be reduced by increased wilting periods during silage making, or by cutting grass with a higher dry matter content, i.e. making haylage rather than silage.

The first element in managing animal wastes is to know how much material you are dealing with, what potential pollution threat it poses and how to minimise the volumes involved, which can be surprisingly large (see Table 7.3).

Waste faecal material mixed with bedding is termed manure or FYM (farm yard manure), which is semi-solid and handled mechanically. Slurry (sometimes termed animal effluent) in contrast is liquid manure (faeces and urine mixed) produced by animals housed without bedding; it has a dry matter content of $30\text{--}120\,\mathrm{kg\ m}^{-3}$ and is handled as a liquid. Both are potentially polluting, but solid manures tend to be easier to process.

Table 7.2. *The polluting potential of different organic waste materials is measured as Biochemical Oxygen Demand (BOD). The higher the BOD the more polluting the waste. When organic material enters a watercourse, it is broken down by micro-organisms. This process uses oxygen that is essential for other stream-life. BOD is measured as the amount of oxygen used by micro-organisms in breaking down a waste material*

Type of farm waste	Biochemical oxygen demand (mg l^{-1} of oxygen)
Cattle slurry	17 000
Pig slurry	25 000
Silage effluent	65 000
Waste milk	100 000
Dirty yard water	1 500
Parlour washings	1 000–2 000
Vegetable washings	500–3 000
Domestic sewage	300

Table 7.3. *Livestock manures are extremely variable, not only in the volumes produced, but also in their chemical, physical and microbial compositions (see Table 7.2)*

Type of livestock	Volume manure produced (litres/day)
1 Dairy cow	35.0–57.0
1 Beef cow	30.0
1 Calf	7.0
1 Dry sow	4.0
1 Lactating sow & litter	14.9
1 Mature sheep	4.0
1 Fattening lamb	2.2
1000 Laying hens	49.0
1000 Pullets	120.0

As can be seen from Table 7.2 even dirty yard water and parlour washings have considerably more polluting potential than does raw human sewage. Thus a key component in managing animal wastes involves reducing the amount of contaminated water produced by minimising the extent of uncovered yard space and ensuring that roof gutters and drainage systems function properly and are separate from slurry holding tanks. Keeping parlour washings and grey water separate from animal wastes reduces the volumes of slurry that need disposal. Another essential element in farm waste management is reducing nutrient losses by reducing the surface areas involved. Nitrogen is lost in the form of ammonia and nitrous oxide via volatilisation from the surface of manures. Therefore reducing the area contaminated with faeces reduces these losses. For this reason, the open surfaces of slurry tanks are best left to form a crust, which reduces nitrogen losses via volatilisation. Crust formation can be encouraged by spreading straw on the slurry surface; alternatively volatilisation losses can be reduced by covering the surface in a layer of vegetable oil. The release of ammonia fumes generated when slurry tank crusts are disturbed can be potentially dangerous. The process of calculating the volume of organic waste that needs dealing with is the first step in farm waste management planning. This is often standardised into a table or spreadsheet format that simply multiplies the numbers of different types of livestock on the farm by the period of time for which they are housed. This is a typically European approach, while in the United States and, to some extent, Canada nutrient management plans identify application limits for nitrogen and phosphorus based on nitrogen and phosphorus stocks in the soil as well as physical land characteristics.

The next stage in farm waste management planning is to calculate how much land is required to be able to spread a known volume of manure, without the risk of polluting a watercourse or a sensitive low nutrient habitat. This value is then compared with the amount of suitable land available to determine if more land or storage facilities are needed. The procedure is again usually standardised into a table format, with the first step being to identify which land is suitable for receiving manure applications. The risk of pollution associated with each area of the farm is identified and typically colour coded as follows:

Red Areas (where slurry should <u>never</u> be spread). Land which is:

> 10 m from a ditch or watercourse
> 50 m from any spring or borehole used by humans or a farm dairy
> On steep slopes and wetlands with high run-off risks
> Other land restricted by tenancy agreement, protected by a
>   conservation designation or agri-environment agreement

Orange Areas (should not ideally receive slurry applications). Land where:

> The soil has been compacted
> There is a risk of flooding
> The soil is waterlogged – at field capacity

White Areas (where slurry is not spread). Land which is:

> Non-farmed, buildings, roads, etc.
> Orchards or woodlands
> Semi-natural habitats, species-rich grasslands, etc.

Fields located too far from the farmstead

Rocky and uneven

Once all the land has been assigned a pollution risk, then the total area identified as being suitable to receive slurry applications can be compared with the volume of slurry available. If there is enough land left, the next problem is identifying an appropriate day and method for applying slurry. Although farmers like to apply slurry on ground that is frozen (because it reduces the risks of soil compaction) this is not advisable as it is likely to run-off and plants are unlikely to be actively growing and thus utilisation rates will be low. Application methods follow the same principle identified above in terms of reducing volatilisation by reducing surface areas involved. Therefore application methods that inject slurry below the soil surface rather than spraying it over the field are associated with reduced nitrogen losses both to the atmosphere and via leaching. In addition, reduced faecal contamination of the pasture increases its attractiveness to grazing animals and reduces problems in silage making.

The farm waste management planning process outlined above is a fairly crude affair designed to prevent pollution incidents and problems like those seen in Figures 7.4 and 7.5. The calculations involved are much more finely refined in nutrient budgeting, which can be performed separately for different nutrients. Spreadsheets exist which enable farmers to balance the nutrients within the farming system with the aim not just of preventing pollution but of saving costs by precisely matching the nutrient requirements of the crops (forage or arable) with application rates of organic and inorganic fertilisers. In nutrient budgeting inputs of various nutrients are calculated separately for artificial fertiliser inputs, effluent (slurry) applications, those available from the soil pool and for nitrogen via fixation as applicable. Outputs are estimated for nutrients exported in the crop as milk, wool or meat, or even tied up in live-weight gains of animals not taken off the farm. Nutrients lost via leaching or those which become unavailable within the soil are also estimated. As with farm waste management planning the final step is comparing the values to determine if the farm is in nutrient surplus or deficit. The results reveal that in intensively managed livestock farms nutrient inputs exceed nutrient outputs, while the opposite is generally true for extensively managed low-input farms. In the past if such nutrient imbalances occurred at all they did so at the small scale, but there is increasing evidence that problems are now occurring at the regional scale.

Livestock production systems in tropical countries can have very different patterns of housing. Typically animals are not housed but grazed at low stocking densities even through dry seasons when forage growth is limited, and therefore waste management is less of an issue. However, in areas with high population densities, for example Caribbean islands, livestock may sometimes be housed at night or for part of the year and fed on cut forage. Under these conditions, the manures produced are treated as a highly valued commodity, composted and sold to support intensive horticultural production.

Microbial pollution

In the past most water quality measures were typically defined in terms of levels of nitrates and phosphates. However, the European Water Framework Directive regulates bathing water quality using faecal indicator organisms as compliance parameters. When this was first introduced it was assumed that untreated human sewage was responsible for observed failures of compliance; however, it quickly became apparent that many incidents of faecal microbial contamination of bathing waters were related to agricultural activity. It has been argued that this is related to the higher concentrations of microbes within ruminant animals; however, it is not yet clear what proportion of these microbes survive in the faeces or how long they remain viable in bathing water, although it is clear that livestock produce considerably more faeces than do humans.

Microbial pollution incidents do not appear to occur at random, but are spasmodic and associated with periods of heavy rain. Attempts to ameliorate microbial pollution from agricultural activities have so far achieved mixed results (Kay et al., 2005) but given the significance of E. coli 0157 and Cryptosporidium for human health, the management of animal wastes is only likely to grow in importance.

Gaseous wastes and greenhouse gases

Until concerns were raised about the greenhouse effect and human-induced climate change, little attention was given to the waste gases produced during agricultural production. This is no longer the case and since agriculture is the second largest global producer of greenhouse-gas (GHG) emissions after industry, there is now considerable research attention focused on the problem, although in the United States and Australia this research may be justified as increasing feed utilisation efficiency, in part because they have not ratified the Kyoto protocol. Of the GHG involved, agriculture is the main producer of methane (25% to 40% of global anthropogenic emissions) and nitrous oxide (40%). Significant amounts of methane are produced by enteric fermentation in livestock, rice cultivation and manure handling, and nitrous oxides are released as a result of inefficient use of nitrogen in agricultural soils, again often associated with poor waste management. In addition agriculture is responsible for producing significant amounts of carbon dioxide (10% to 30%) by the burning of crop residues, from decomposition within the soil and from energy consumption both on- and off-farm. Although methane is present in the atmosphere at lower concentrations than is carbon dioxide (1.7 ppm as opposed to 380 ppm) it is more damaging as a greenhouse gas, although it has a shorter lifespan in the atmosphere, surviving for only 12 years while carbon dioxide may survive for up to 200 years. Methane has a carbon dioxide equivalent value of 21 while for nitrous oxide it is 310. This means that the global warming potential of one tonne of methane and nitrous oxide is equivalent to 21 and 310 tonnes of carbon dioxide respectively.

As most of the nitrous oxide produced by agriculture is the result of poor nutrient management, there is some reason for believing that these emissions can be reduced. This is firstly because a number of appropriate management practices are already well established (see above) and secondly, economic savings which can result from this (reduced requirements to purchase fertiliser, avoiding prosecution for polluting) accrue directly to the farmer, rather than to the global population, hence the parable of the commons does not apply. In contrast reducing methane production from livestock seems more difficult. With the average lactating cow producing 200 to 400 litres of methane a day, combined with a global population of 1 200 000 000 domestic cattle the result is a considerable impact on the climate. Attempts to reduce methane production include altering rumen microbial populations either genetically or by manipulating the diet, or by livestock breeding. Balancing the nutritional requirements of livestock with their feeds can result in reductions in nitrous oxide and methane production. As with better manure management, improvements in the diet can result in savings to the farmer, as well as the environment, but so far this seems a more difficult message to get across as the benefits tend to be less direct. Therefore, the most effective way of reducing agricultural GHG emission may be reducing the numbers of domestic animals, but this seems unlikely given a rising human population with an increasing appetite for meat.

Fallen stock: wildlife resource or disease reservoir?

Dead animals or fallen stock are one of the more unpleasant wastes unintentionally produced by agricultural activity. The issues surrounding the disposal of carcases are typical of those of many farm wastes. Over time, there has been increasing legislative control of the disposal of fallen stock, driven by concerns over the spread of diseases and pollution of water tables. There has been a shift away from informal on-farm disposal, by small-scale burial or in the United Kingdom feeding to local packs of fox hounds. These methods have tended to be replaced by more regulated incineration within tightly defined periods of death. In many remote hill farming areas, it is very difficult to ensure that such time limits are adhered to, and furthermore the collection and removal of carcases from mountainous areas can be challenging. It has been argued by some conservationists that a few dead hill sheep represent a minimal pollution risk, while their removal may deprive raptors and other scavenging carnivores of a valuable food resource. In fact the removal of fallen stock may force carnivores into taking live animals, turning them from a useful garbage disposal service into conflict with the farming community. Is this an example of working with nature rather than against it, as is often claimed by organic farmers? In a related issue, declines in vulture populations across the Asian tropics are thought to have resulted from birds being poisoned by unregulated veterinary medicines that are found in the carcases on which they feed. The appropriate and speedy disposal of fallen stock is particularly important in tropical agriculture. Although regulations may be limited, obvious health concerns and avoiding noxious smells result in on-farm disposal of carcases by burning or burial being rapid.

Outbreaks of disease amongst livestock, such as bovine spongiform encephalopathy (BSE), or foot and mouth disease, which occurred in 2001 in the United Kingdom, can present great logistical problems for disposing of carcases. With BSE it is considered essential that culled animals are incinerated at the high temperatures required to denature prions, the infective agent. This requires the use of specialist incinerator facilities which may have limited throughput. In contrast the disposal of large numbers of carcases associated with an outbreak of foot and mouth tends to involve the use of large pyres or burial pits. Both of these alternatives are associated with potential pollution risks, either atmospheric or waterborne.

Farm waste material imported on to the farm

The list of materials that may be imported onto a modern farm is vast and diverse and many of these eventually end up as wastes. Unlike waste products produced on-farm imported wastes tend to be inorganic in nature or complex potentially highly toxic chemicals. The main categories of solid wastes are: packaging, pesticide containers and silage wraps, scrap machinery, batteries, tyres, building wastes and old fencing materials. In the past the solid wastes were typically buried in a remote area of the farm, dumped down old mine shafts, abandoned in an unsightly heap in the farmyard, or incinerated.

Such practices are still commonplace in many countries; however global environmental legislation regulating their disposal is becoming more stringent. The alternatives to on-farm disposal are off-farm specialist waste disposal or recycling companies, but because of the logistics involved in collecting wastes from remote rural areas, such services tend to be expensive. If governments are serious about addressing this problem, then perhaps there is a case for subsidising farm waste disposal rather than relying entirely on a legislative solution. This approach is likely to be more effective, because the small-scale diffuse pollution incidents that arise from unregulated illegal disposal on-farm are very difficult to trace and prosecute. Alternatively there have been suggestions that manufacturing companies should be responsible for the subsequent disposal of their products and spent packaging, after use.

Imported liquid wastes form an equally diverse group including: contaminated oils, leftover pesticides, spent sheep-dips and old veterinary products. As with solid wastes, these have generally been disposed of on-farm, by dilution and spraying on to land away from watercourses and at low risk of run-off, or again by incineration. Here too there is a movement towards specialist off-farm disposal, driven by legislation. As described in Chapter 3, some agrochemicals have the potential to do great environmental damage and although there has been a move towards more benign chemicals with shorter half-lives, the move from organophosphates to synthetic pyrethroid sheep-dips proves this trend is not a rule. Although synthetic pyrethroids are less damaging to human health, traces of them have been responsible for killing invertebrates in many miles of streams and it is therefore appropriate that their disposal is tightly regulated.

Given the expense of off-farm disposal of imported farm wastes, the '3Rs' mantra of the green movement should be considered: Reduce, Reuse, Recycle. Of these, recycling suffers the same problem of costly collection costs in rural areas as disposal, and there are limited possibilities for on-farm reuse that have not been exploited for years by prudent farmers. This leaves reducing waste production as the most likely avenue for successfully reducing the costs of specialist waste disposal.

In addition to waste products generated by agricultural activity there is a further set of wastes that need consideration, those that are generated off-farm but may be disposed of on-farm as a potential nutrient source. These include human sewage and organic industrial wastes such as paper pulp and abattoir wastes. There is an ecological logic to disposing human wastes to land in terms of closing nutrient cycles, and in the past this practice was widespread. However, primarily because of concerns about disease spread this is illegal in many regions. In fact it should be relatively simple to ensure that human effluents returned to land are free from disease-causing agents. A more serious agricultural concern may be that of heavy metal contamination. The most problematic metals are copper, zinc and cadmium, which do not come from domestic sewage but are largely associated with industrial waste and therefore the problem should be solvable. Unfortunately the levels of metal contamination currently found in most human effluent can only be spread on agricultural land for 20 to 30 years before toxic effects are found to inhibit crop growth. Abattoir wastes are similar to human wastes in rising health concerns, and again can be seen as partially closing nutrient cycles. As yet the disposal of industrial wastes to farmed land is not widespread and its long-term environmental impacts are uncertain.

Summary

We have seen in this chapter that a wide range of waste materials are found within the farmed environment. Typically those produced on-farm (although they have the potential to be highly polluting) are best regarded as nutrient pools and managed as a resource rather than as a problem. There are now well-established codes of good practice and software packages available to assist farmers in managing the nutrients found in organic farm wastes. Effective management of animal wastes not only benefits the environment, but also reduces farm costs. In contrast waste materials that are generated on-farm from imported products are generally of little value and indeed can represent a considerable cost to the farmer, who is increasingly required to seek licensed contractors to dispose of such materials. Farmers and the rest of the human population as consumers have to accept responsibility for producing too much waste for too long.

In some countries an increasingly important driver in changing the way that farms deal with waste is the increasing concern of farmers about the risk of being legally liable for environmental offences. With greater public awareness and larger non-farm rural populations the potential for conflict over environmental issues has increased. Some areas have implemented 'right-to-farm' legislation to protect farmers from nuisance lawsuits (e.g. smells and noise in farming areas). Environmental farm plans are being used to identify environmental problems and focus environmental risk management as well as helping to mitigate liability over environmental damage.

8

Low-impact farming systems

Introduction

This chapter is entitled 'low-impact farming systems' by which we mean agricultural systems designed (at least in part) to have less dramatic impacts on the environment than does conventional agriculture. There are however a number of problems with defining low-impact farming systems in this way. Firstly, what are environmental impacts, how do you measure them and indeed can environmental change be simply measured on a scale of good through to bad? Secondly, our definition is a comparative one, measured against conventional agriculture, but this begs the obvious question – what is conventional agriculture? Thirdly, what is a system, how tightly restricted are the various methods of farming? In this chapter we examine these issues in some detail, because unless we can rigorously define what a low-impact farming system is, then how can we design one or measure what its environmental impacts are? This chapter reviews the aims and methods of some of the more common farming systems that claim to have environmental benefits or to be less environmentally harmful, including organic agriculture and agri-environment schemes. The theory and practice of sustainable agricultural production is discussed along with the considerable technical difficulties involved in comparing the environmental impacts of different production systems. The assessment of relative environmental impacts of different farming systems is not only difficult; it is central to being able to design low-impact farming systems, because how do you know if your aims are being achieved unless you can quantify them?

What are environmental impacts and how are they assessed?

Agricultural activity changes the environment in many ways and at many scales. These include abiotic factors such as water and air chemistry and

143

biotic factors such as the balance of species present at a site. Some of these changes are easy and inexpensive to measure accurately, while others less so. Agriculturally induced environmental changes can occur at a local scale, for example noise, at a regional level such as eutrophication or changes may be global such as contributing to climate change. Different agricultural activities will of course result in different suites of effects. As a result of these complexities, there is no standard measure of environmental impact by which agricultural systems can be compared.

Methods for estimating environmental impacts at the farm level include input–output accounting (IOA) sometimes called green accounting (Halberg *et al.*, 2005). The approaches used in IOA vary but typically are based on somewhat subjectively selected 'good agricultural practices' (GAP) and indicators that are easy to calculate such as the amount of pesticide use per hectare, farm-gate nutrient balances or energy use per kilogram of product. Other farm-level measures are based on predicted emissions or nitrogen losses or ecological footprint analysis. It has long been appreciated that how IOA is performed can radically affect the outcome of comparative studies. Lampkin (1997) compared organic and conventional dairy farms in terms of their energy budgets. Per unit area the organic dairy was found to use less energy because of reduced demands for diesel and electricity. But when this comparison is made per litre of milk produced, the advantage swings in favour of the conventional farm with its higher productivity. If the comparison is extended to include the off-farm energy inputs such as that used in the manufacture of fertilisers and pesticides, then again the organic system scores better. However, if post-farm food miles are added into the equation then the balance may move again. It can be argued that each of these refinements to the environmental assessment makes them more realistic, but it is also more problematic to collect the required data. Lampkin (1997) also considers other aspects of sustainability, including food supply and security, financial viability and social impacts; with each additional factor included identifying the correct balance to generate an overall farm assessment becomes more difficult. Extending the number of factors considered better represents the multifunctional nature of the farming enterprise, but this appears to be traded-off against objectivity.

A potentially simpler alternative approach has emerged from attempts to measure the environmental impacts of genetically modified crops. This can be termed solar energy accounting, and depends on estimating the proportion of solar energy intercepted by a crop and diverted into human uses versus the proportion of solar energy available to support other species. The strength of this method is that it provides a conceptually simple way of measuring the efficiency of agricultural production, or conversely its environmental impacts.

Unfortunately, estimating the energy content of total on-farm biodiversity is practically more difficult, and this method fails to account for other environmental impacts such as water and air quality or fossil fuel use etc.

Other methodologies have been developed to estimate the environmental impacts that arise from farming at a regional level. These include: environmental impact assessment, agri-environmental indicators, multi-agent systems, environmental risk mapping, life-cycle analysis and linear programming; these have been reviewed by Payraudeau and Van der Werf (2005).

Environmental impact assessment is a tool used to predict the environmental impacts of a change in land management, as well as identifying ways to mitigate any predicted adverse impacts. It aims to design projects that match their local environment and present predictions and options. As such it typically produces rather general guidance rather than a precise scientific measure of environmental change.

Environmental indicators are variables that are selected as being statistically robust and easy to measure (see Chapter 4 for a discussion of agri-environmental indicators). Typically six principal indicators are used: soil phosphorus supply, potential nitrogen loss, risk of soil erosion, risk of polluting the water supply, energy use efficiency and potential impact of farmland biodiversity. In practice these are often used as measures of pollution per unit of production.

Multi-agent systems models define within a single model many separate entities or agents, which interact with their environment and, via modifying the environment, with each other. These agents are defined at different hierarchical levels and can include not only agricultural managements, but socio-economic factors that influence them, within a complex network of interactions of which environmental changes are just one element.

Environmental risk mapping is a tool that is rarely used, and like environmental impact assessment is generally used in the planning process. As its name suggests it involves a spatially explicit analysis of land management options in relation to potential environmental threats that may result.

Life-cycle analysis is a technique that assesses the environmental impacts throughout the entire farming system. Each step of the process is considered including the inputs of materials, on-farm activities plus the use of the end-products and disposal of waste materials. Outputs can be expressed per unit of production or land used, reflecting the dual nature of farming, production and occupancy of land.

Linear programming is a system of modelling that is used to predict the impacts of various agricultural scenarios. It assumes a level of mathematical understanding of the process at work which links the changes in land management with their resulting environmental impacts. This kind of modelling is improved as understanding of the systems advances and reliable information becomes available. The strength of the approach lies in its use of spatial information for assessing agricultural land use options within the context of regional planning. Its weakness lies in our limited understanding of many of the mechanisms involved.

The complexity of the assessment methods outlined above in part explains why no standards have been adopted. In the view of Payraudeau and Van der Werf (2005) better indicators are ones that incorporate interactions and uncertainty. These uncertainties should be based on environmental effects rather than inputs or practices. Better indicators should produce predictions that can readily be validated and easily understood. It seems unlikely that a perfect solution exists.

There are a number of fundamental problems that must be addressed when trying to design a method of assessing environmental impacts. Classical ecological succession theory predicts that if you manage (farm) any piece of land in a constant way for long enough (and this can be very long) then a stable equilibrium community of plants and animals should develop. If this was the case then the environmental impacts of a particular farming system could be assessed by comparing its equilibrium state with a similar but non-managed area of land or land farmed in a standard form. However, this simple theoretical equilibrium state is unlikely to develop under an agricultural regime for a number of reasons. It is now difficult to conceive of a farming system that remains constant over a long enough period of time for a stable plant community to develop. Changes in economics and technology mean that agricultural practices change frequently so that the environmental impacts that result are also dynamic. In the natural unmanaged state, soil nutrients and organic matter tend to increase slowly during succession. In the farmed environment nutrients may be imported as artificial inputs or exported as product. There are very few long-term studies of biodiversity on land under constant human management. Perhaps the best known example is the Park Grass Experiment at Rothamsted (see Figure 8.1).

The Park Grass Experiment illustrates another problem that needs addressing when assessing the environmental impacts of any agricultural practice. It may seem reasonable to assume that comparative indexes are fixed against a constant environmental standard. But what system can be used as an environmental standard when we know that land that has been managed for more than 100 years in a fixed way is still changing for reasons that are not fully understood? Such change may be driven by external non-agricultural factors such as climate change (see Figure 8.2) but this again only highlights the problem of ascribing environmental change to an agricultural practice when other

The Park Grass Experiment

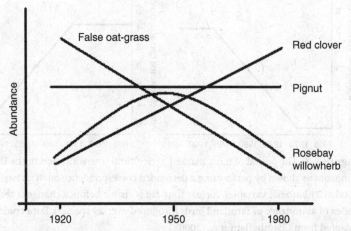

Figure 8.1 Long-term data from Rothamsted demonstrate that biodiversity change occurs even in the absence of changes in agricultural practice. Data crudely adapted from Dodd *et al.* (1995).

Mean maximum temperature for July & August at Rothamsted

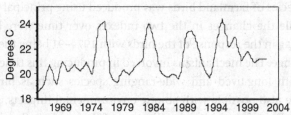

Figure 8.2 Weather records from Rothamsted indicate rising summer temperatures. Such external drives for environmental change make the task of assessing the extent of agriculturally induced change more difficult. Data derived from Rothamsted Meteorological Station.

unknown environmental factors may also be varying and interacting with agricultural practices.

The difficulties in comparing the environmental impact of different agricultural systems relating to scale are discussed in Chapter 9. But with many environmental impacts resulting from changes in agricultural activity there can also be the related complication of a time lag. This has been clearly demonstrated in the case of the decline in farmland birds in the United Kingdom.

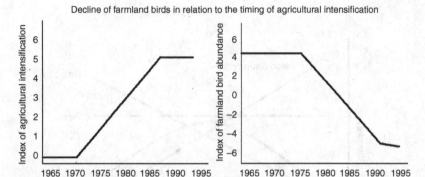

Figure 8.3 The period of most marked agricultural intensification in the United Kingdom as shown by performing a detrended correspondence analysis based on a total of 31 land-use variables. A clear time lag is visible before a change is seen in the index of abundance of farmland birds (produced over 29 species). Data crudely adapted from Chamberlain *et al.* (2000).

Chamberlain *et al.* (2000) produced an index of agricultural intensification using detrended correspondence analysis, based on a total of 31 land-use variables (the majority of which were highly intercorrelated). Plotting this index over time revealed that 1970–88 was the most intense period for agricultural intensification in England and Wales (see Figure 8.3). A comparative index based on the abundance of 29 species of farmland birds was produced using principal components analysis. While the changes in the two indexes over time were broadly similar there was a lag in the response of the birds with 1974–91 being the period of most marked change. The mechanisms involved in producing this time lag are easy to imagine with long-lived and wide-ranging species such as birds, but similar time lags are also possible with abiotic environmental impacts, such as the eutrophication of aquifers. A clear implication of this study is that it is not possible to identify individual agricultural activities responsible for many environmental impacts without resorting to a detailed experimental approach.

Why do environmental impacts matter and what environment do we want?

Having considered the range of environmental variables that can be affected by agricultural activity and the difficulties involved in measuring these impacts, we must ask – why should we be concerned about environmental impacts? Of course agriculture changes the environment, that is what agriculture is, changing the environment to benefit species that humans wish to exploit. This question is similar to the generic question of why humans should conserve/protect their environment that has been considered many times before. Norton (1987) describes a taxonomy of rationales for why humans should preserve natural variety which is readily adapted to the agricultural situation.

Economic reasons

1. Anthropocentric reasons – at the simplest and most self-interested level there are direct economic reasons for farmers being interested in protecting the environment. In the days of agri-environment schemes farms receive payments specifically for looking after the environment. Although some farmers see their job as a way of life, any agro-ecologist who forgets that farms are primarily businesses with farmers making management decisions based on economic signals is in danger of misunderstanding their function.

2. Non-anthropocentric reasons (or cases where the economic benefits are indirect) – these reasons are typically referred to as maintaining ecosystem services, or maintaining the environment in a state that is able to meet the requirements of humans and other species. These requirements include: a supply of pure water and air adequately free from pollution, a climate free from excessive extreme events, functioning nutrient cycles and soil processes, etc. When any of these factors are damaged then economic costs are incurred, but as discussed in Chapter 4 these are often external costs that are not always easy to ascribe to a specific damage-causing agent.

Ethical reasons

1. Moral reasons – these may emerge from religious beliefs, but are by no means exclusive to religion. There is a strong weight of opinion that humans in general and agriculturalists in particular are custodians of the environment. This argument states that as land managers we have a moral obligation to protect the other species with which we share our planet and the environment upon which we all depend.

2. Aesthetic reasons or anthropocentric ethical reasons – these arguments state that we should protect our environment because we gain enjoyment and pleasure from it. In a developed world agricultural context, this could be extended to: many consumers now consider that the products of farming include countryside as well as wholesome food. This logic may return us to the first economic reason. There may be direct economic benefit to the farmer if not in considering the

environment as a product, but in being aware that his/her food products may be more marketable if they are associated with an aesthetically pleasing mode of production. That is the consumer may not directly pay for the aesthetic value of the countryside, but may be more willing to purchase food that has been produced in an aesthetically pleasing landscape. This is a phenomenon captured by eco-labelling initiatives.

Understanding why we wish to protect our environment is a prerequisite to knowing what it is we wish to protect and how to achieve this. For example if our rational is primarily driven by non-anthropocentric arguments about protecting ecosystem functions, then our priorities should be driven by ecological understanding. However, if our reasons are heavily influenced by aesthetic values, then our actions may be more open to the whims of fashion. In reality environmental protection is driven by a range of different motivations, which are rarely questioned in these terms. However, a significant element of human aesthetic preference can be detected and it is difficult to defend in a rational way. For example, superficially Chamberlain et al.'s (2000) index of change in abundance of farm birds (Figure 8.3) implies a decline in farm birds with agricultural intensification. The observed change in abundance of farm birds is generally regarded as being undesirable. However, in fact the index incorporates increases in numbers of jackdaws, rooks and stock doves. This begs the question – why the concern? Does it matter if partridges, lapwings and skylarks have been declining if they have been replaced by other species? The ecosystem seems to have moved from an equilibrium state associated with traditional agriculture to another associated with modern agriculture. How should we determine what is the correct balance of nature? What is the right ratio of skylarks to jackdaws? It has been argued by Warren (1995) that the public consider ecological communities associated with former agricultural practices to be more aesthetically desirable than those associated with current agricultural practices, and in the future they may look back nostalgically on the beautiful fields of set-aside of the 1990s. It would seem more scientifically defendable to define desirable environmental conditions based on sound ecology than human prejudice, but this debate is rarely had. This is also regionally variable, for example in North America where the history of agricultural development is much shorter and pristine, or at least minimally impacted landscapes do exist, environmental objectives are often met on agricultural landscapes using land idling and management schemes that provide areas that resemble pristine areas. In these regions then the benchmark landscape that influences the objectives of agri-environmental policy is not, in general, one that includes agriculture. A perhaps extreme example of this is the debate on re-establishing the 'Buffalo Commons' in the short grass prairie of the great plains of North America. It is argued that the current depopulation of this area is evidence of a failed development expansion experiment with the area being unsuitable for agriculture. Therefore, the Buffalo Commons is proposed as an alternative whereby large parts of the Plains would be restored to their pre-European contact condition to make them again the commons the settlers found in the nineteenth century. 'By creating the Buffalo Commons, the federal government will, however belatedly, turn the social costs of space – the curse of the short grass immensity – to more social benefit than the unsuccessfully privatized Plains have ever offered' (Popper and Popper, 1987).

How are low-impact farming systems defined?

As we have seen there are several motivations for wanting to protect the environment and many ways of assessing if environmental change is occurring. These combined with different historical factors have resulted in a host of different agricultural systems which make some claims to be better for the environment (Thirsk, 1997). The socio-histories of some low-impact farming systems are perhaps surprising and complex, but these are out of the scope of this book. However, it should be noted that several apparently opposing factors have sometimes contributed to the development of low-impact farming methods. For example, Integrated Crop Management can be said to have two separate evolutionary lineages. Firstly, following the landmark publication of Rachel Carson's *Silent Spring* in 1962, there was public pressure for agriculture to become more environmentally benign. This was partially responsible for the movement away from generalist pesticides such as DDT and favoured the development of more chemically and mechanically targeted products. The increased cost of production of second and third generation pesticides acted as another driver (this time acting on the industry) which also favoured the reduced use of chemical inputs. Both of these and other pressures were important in shaping the aims and practices of low-impact farming systems. This background is needed if we are to understand how various low-impact farming systems define themselves and their practices.

Organic agriculture

Although what constitutes organic production is legally defined in many countries, historically the organic movement has been divided into many accreditation organisations, which have differed subtly in their definitions and there is still no single approved definition of organic agriculture. Indeed organic growers can be divided into those inside and outside of 'the moment'. Use of the word organic in relation to agricultural production can be traced to Lord Northbourne (1940) in the United Kingdom. The Soil Association was founded shortly after in 1946 to 'research, develop and promote sustainable relationships between the soil, plants, animals, people and the biosphere'. The Soil Association in the United Kingdom now defines the aims of organic farmers as 'to produce good food from a balanced living soil. Strict regulations, known as standards, define what they can and can't do. They place strong emphasis on protecting the environment' (Soil Association, 2006). In the United States the USDA (2006b) define organic production as 'A production system that is managed in accordance with the Act and regulations in this part to respond to site-specific conditions by integrating cultural, biological and mechanical practices that foster cycling of resources, promote ecological balance, and conserve biodiversity'.

These two definitions are typical of organic agriculture, in that they encompass the multifunctional nature of agriculture, incorporating environmental impacts, food quality, cultural and economic aspects. These definitions also have a strong element of what organic farmers 'can and can't do'. This has arisen by necessity, to give the public confidence in organic products by preventing farmers exploiting the market by claiming to be organic without actually being so. Definitions of organic agriculture also tend to include statements about protecting the environment, and maintaining or promoting ecological balance. These aspects are, however, imprecisely defined and it is not clear if the ecological balance refers to an element of ecosystem functioning or if it is more an aesthetic balance such as skylarks being preferred over jackdaws. However, the strong emphasis on soil processes and soil health within these definitions tends to imply that the environmental aims of organic agriculture are related to maintaining ecosystem services rather than some more idyllic vision of the countryside. The organic movement's aims of enhancing food quality and the environment have perhaps focused more on the environment in recent years, as it has proved difficult to demonstrate any health benefits from organic produce.

Sustainable agriculture

The history of sustainable agriculture is closely allied with that of the organic movement and finding a universally accepted definition for it is famously difficult. Discussion of the topic is fraught with unscientific comments such as 'sustainability is a direction rather than a destination' and 'sustainability is a question rather than an answer'. The problems defining sustainable agriculture appear to result from the word sustain, because it is not clear what is being sustained or for how long. Available definitions tend to include the environment as one of the factors in need of sustaining, but it is less clear in what state it should be sustained. There is also some general agreement that sustainability should be maintained permanently or at least for a long time, which of course is difficult to verify. In fact, whether a system is sustainable or not is impossible to determine until some point in the future when it can be evaluated if, up to that point, the system was sustainable.

Published definitions of sustainable agriculture are similar to those of organic agriculture in encompassing environmental protection amongst a list of other important factors. Definitions are typically vague and lack the regulatory restrictions of can and cannots associated with organic farming. Typical examples include: in general, sustainable agriculture addresses the ecological, economic and social aspects of agriculture. To be sustainable, agriculture can operate only when the environment, its caretakers and surrounding communities are healthy (Leopold Centre, 2006) and a sustainable agriculture must be ecologically sound, economically viable and socially responsible (Ikerd, 2006). As such, sustainable agriculture does not explicitly preclude the use of artificial inputs in attaining its goal of sustainability, but such practices are generally disapproved of. The motivations for protecting the environment and methods employed are therefore similar to those of organic farming and can be said to relate to ensuring working ecosystem services, with the target environment being defined vaguely as one able to deliver these.

An operational definition of sustainability that has emerged is based on the characterisation of a system (including agri-environmental systems) as being comprised of a collection of capital stocks. One of the most common categorisations of capital stocks includes natural capital (biological based resources), man-made capital (equipment and buildings), human capital (management expertise, labour) and social capital (institutions and relationships). The capital stocks are combined for the system to be productive. Based on this characterisation, the productive system is considered sustainable if the capital stocks are not eroded or degraded over time. That is, the capital stocks are conserved such that the productivity of the system is not increasingly constrained by the quantity and/or quality of a given capital stock. There is, in fact, a continuum of sustainability interpretations from weak sustainability to strong sustainability. Weak sustainability requires that the overall stock of capital should remain constant (Common and Perrings, 1992). Weak sustainability enables the reduction of a capital stock as long as another capital stock is increased to compensate (e.g. increased investment in chemical nitrogen to compensate for losses of natural soil nitrogen). Strong sustainability, in contrast, requires that each of the capital stocks must be kept constant, with a particular emphasis on natural capital (Common and Perrings, 1992). Strong sustainability assumes that natural capital and other forms of capital cannot be substituted and that there is inherent uncertainty and irreversibility associated with natural stock degradation.

At the simplest level an assessment of whether the system is sustainable, then, can be operationalised into an assessment of the relative changes in the quantity and/or quality of the relevant capital stocks. If, for example, it is found that the population of insect pollinators has decreased within an agri-environmental system then this system may not be sustainable. However, there are obvious difficulties in this assessment including the identification of a baseline or benchmark capital stock, selection of appropriate physical indicators to quantify changes in capital stocks and the accurate valuation of capital stocks to enable accurate capital stock substitution.

Permaculture

Permaculture is more than a system of agricultural production; the concept encompasses an entire way of sustainable living. The term was first coined in 1976 by Bill Mollison and David Holmgren and expanded upon in 1978 in their book *Permaculture One*. Definitions of permaculture struggle with the same problems as those facing sustainable agriculture: what is being sustained, in what state is it being sustained and for how long? For example, the Permaculture Net (2006) define permaculture as an agro-ecosystem that is designed and maintained by its owner/occupiers to provide for their food, energy, shelter and other material and non-material needs in a sustainable manner. It appears that the intention is the sustainable provision of all human needs, but less clear is what environmental conditions are acceptable in attaining this. Other definitions like the following one from Permaculture International (2006) incorporate elements that have echoes of ecosystem processes 'consciously designed landscapes which mimic the patterns and relationships found in nature while yielding an abundance of food, fibre and energy for provision of local needs'. By implication the objectives of permaculture must include both utilising and maintaining ecosystem services however vaguely they are defined and as such they are again similar to those of organic farming.

Biodynamic agriculture

Biodynamic agriculture is regarded by many scientists as the embarrassing lunatic fringe of the organic movement. Perhaps as a reaction to this, biodynamic agriculturalists make more claims about being scientific than do those defining other low-impact farming systems. Their literature frequently talks about the science of life-forces, but these are not forces that most scientists would recognise that can be measured in newtons. The principles of biodynamic agriculture are derived from Rudolf Steiner (1861–1925) and his spiritual philosophy known as 'anthroposophy'. The approach is based on integrating observations of natural phenomena with knowledge of the spirit. In practice biodynamic farming has many practices in common with organic systems, but it also incorporates spiritual elements, which although they are scientifically testable have been more likely to attract ridicule than serious scientific consideration. Elements of the biodynamic philosophy which talk about the spiritual history of the Earth as a living being have some parallels with James Lovelock's Gaia hypothesis. While both Gaia and biodynamics have received a sceptical reaction from conventional scientists, it could be argued that in essence what they are describing is what ecologists term ecosystem services. Even so, it is difficult to clarify what are the exact environmental aims of biodynamic farming. The Biodynamic Farming Association (2006) talks about balance and healing without defining what is being balanced or healed; it also describes biodynamics as an ongoing path of knowledge rather than an assemblage of methods and techniques, which is rather different from most working definitions of organic agriculture. It might not be clear exactly what environment biodynamic farming wants, but it appears that its motivations are driven by ethics and spirituality.

Agri-environment schemes

Agri-environment schemes are government programmes set up to financially encourage farmers to manage their land in an environmentally friendly way. The history, motivations and practice of these schemes are covered in more depth in Chapters 2 and 4. Unlike other low-impact farming systems the aims of agri-environment schemes are easy to identify as they are usually stated in the opening pages of the scheme documentation. However, the language used is often as vague as that defining other low-impact farming systems, with typical objectives being defined as protecting and enhancing the countryside. However, agri-environment schemes are often linked to the delivery of national targets for conservation and environmental quality and these are increasingly tightly defined for example in terms of areas of land to be managed, precise population targets for some species or limits on chemicals found in water. The mechanisms employed in meeting these targets are also increasingly based on the results of scientific studies. However agri-environment scheme management prescriptions also need to be acceptable to the wider agricultural community if they are to be taken up, and be bureaucratically policeable if they are to be enforced. Such prescriptions may incorporate some organic aspects but schemes are more focused on achieving their objectives than they are concerned about the process.

Integrated crop management

Integrated crop management (ICM) is an ill-defined set of agricultural practices, with many different names and acronyms. It has a very different background to most other low-impact farming systems, in that it is often supported and promoted by the agrochemical industry and major food retailing companies. Its origins are more mainstream than other low-impact systems, with one of its main aims being optimising the use of expensive agrochemicals in terms of profits, rather than relying on precautionary blanket applications. This optimisation process is built upon a scientific understanding of the ecology of farmed land and the application of modern technologies and methods to achieve its aims. Precisely what technologies and methods are employed is less important than is achieving this aim. The underlying principle is that avoiding waste is good business sense and also good for the environment. It may be argued that any environmental benefits are secondary and serendipitous. However the stated aims of ICM often explicitly include running a profitable business with responsibility and sensitivity to the environment. As with most low-impact systems exactly what these environmental aims mean in any measurable sense is poorly defined. But because ICM involves the utilisation of ecological services for example in terms of natural pest control, by implication its aims must include the maintenance of these processes.

Balancing aims and methods

We have seen in the above review of the main forms of low-impact farming systems that it is difficult to find simple agreed definitions of what they are and what their aims are. Furthermore, if we are to compare them with 'conventional agriculture' (CA) we face the problem that CA is a catch-all term for many different types of farming and, not being formally organised, it is even more difficult to find a definition for, and has no stated aims.

For many low-impact systems definitions of desired environment quality form only one aspect of a multifaceted set of objectives. In virtually every case where environmental aims are explicitly stated, they are imprecisely defined and not easily open to scientific measurement to establish if the aims are being achieved or not. In sharp contrast the methods approved by different low-impact farming systems are typically very tightly defined and much energy is put into discussing and agreeing 'the approved book of rules'. This may be the result of necessity to ensure that standards are being maintained, rules are practical to enforce and to ensure consumer confidence in the system. However, scientifically one might question if the rules governing approved methodologies have become more important than the aims, which because they are usually ill-defined are allowed to slip. Arguably the one exception to this is ICM in which the aims (however poorly defined) seem to be more important than the methodology applied.

Turning the clock back or applying new technologies

We have established that the environmental aims of most low-impact farming systems are generally imprecisely defined. But it is safe to assume that if they could achieve the rural environment of 70 plus years ago and maintain the food production levels (food quality aside) of today then virtually all low-impact farming systems would consider that their aims had been achieved. Theoretically the simplest way of achieving the rural environmental utopia of 70 years ago would be to reinstate all the agricultural practices of the period, and it can be argued that many of the practices associated with low-impact farming are indeed exactly that. However, many low-impact farming systems are quick to state that they are not about turning the clock back and that their methods are built on new scientific understanding of farmland ecology. Even so, an obvious place to start if you were to design a low-impact farming system from scratch would be by reviewing all the changes that have occurred within agriculture within the last century or so.

Although Table 8.1 cannot be derived with much rigour (because it describes the generalities of low-impact farming systems), it does illustrate that many (possibly the majority) of the changes in agricultural practice that have occurred in the last century are as frequently found in low-impact farming systems as they are in conventional farming. For example, the use of mechanisation, which results in much faster and efficient cultivation and harvesting etc., is not precluded from any low-impact system and this change alone is likely to have resulted in huge ecological changes. This lack of differences in many aspects of management is responsible for the fact that the environmental benefits of low-impact farming compared with CA are subtle and often difficult to quantify.

The decision to incorporate the reversal of any agricultural practice listed in Table 8.1 into the 'approved list' for a low-impact farming system will in theory result in a shift in the balance of environmental gain against loss in production. It would therefore be very helpful if the management changes in Table 8.1 could be ranked in terms of the extent of their associated environmental impacts rather than being listed alphabetically. However, the difficulties discussed at the start of this chapter in defining 'what are environmental impacts and how are they assessed' make this task very difficult. Some of the changes are likely to result in declines in water quality; others may result in declines in bird populations or contribute to climate change. Since these impacts cannot be meaningfully listed on a linear scale of environmental damage, they may be better considered in the context of the stated aims of the particular farming system, but as we have seen these tend to be so vague that they are little help in scientifically selecting approved management options.

Table 8.1. *Alphabetical list of the major changes in agricultural practices that have occurred in last 100 years that are likely to have produced environmental impacts.*

Agricultural practice	Occurrence
Abandonment of traditional rotations and crops	+
Agricultural improvement – reseeding and land reclamation	0
Application of lime – increased	0
Biological control – increased	–
Bulk handling of slurry and silage effluents	+
Change from hay to silage	0
Change to winter cereals from spring cereals	+
Changes in burning practices	0
Changes in food supply chain and abattoirs	0
Cultivation of natural habitats	0
Disposal of fallen stock regulations	0
Efficiency of mechanisation and bulk transport	0
Escape of introduced farmed animals	0
Field drains – increased	0
Increase in field size (scale effects/isolation)	0
Increase in stock numbers/stocking rates	+
Introduced species = introduction of diseases	0
Introduction of environmental schemes	–
Introduction of modern breeds/varieties	+
Loss of traditional buildings	0
Mixed farming – loss of	+
New crop genotypes = escape of alien genes and GMO	+
Night-time cultivations	+
Pesticides – increased	+
Piping, dredging or canalisation of streams	0
Seed cleaning	0
Set-aside – introduction of	0
Speed of mechanisation	0
Subsidy systems	0
Use of concentrate feeds	+
Use of fertilisers/production/transport – increased	+
Use of veterinary medicines, drugs and drenches	+
Waste disposal	+
Water extraction – increased	0

+ indicates practices most likely to be associated with conventional agriculture, – indicates the practices more likely to be associated with low-impact farming systems and 0 indicates that there is no great difference between different farming systems, GMO = genetically modified organism.

In addition, over the last 70 or so years the changes listed in Table 8.1 have occurred in parallel in most of the world's agricultural systems. During this process of agricultural intensification, no experimental farms or plots were established in which the environmental impacts of each of these changes could be investigated. This would have required that each change was implemented in isolation. Even if this had occurred, there would have undoubtedly been interactions between factors and scale and spatial effects of the kinds discussed in Chapter 9, which mean that determining the relative significance of each change is virtually impossible. Furthermore this logic is built upon the assumption that a reversal of the agricultural practice will result in a return to the previous environmental conditions. This may be true in some cases, but in many (for example when the practice has already resulted in species extinction) former environmental conditions may never be achieved. However, before we abandon hope to this counsel of despair, the computer-based landscape modelling methods outlined in Chapter 9 do now offer a rational scientific method for designing/refining suites of agricultural practices to produce a more acceptable balance of production and environmental change.

The effects of organic agriculture on biodiversity

The absence of a single method of assessing the full range of environmental impacts of any particular farming system has limited our ability to carry out robust comparative studies. However, many researchers have investigated the impacts of organic agriculture on farmland biodiversity. These studies have been surrounded by a degree of controversy because different groups of species have been found to respond differently under organic management. Bengtsson *et al.* (2005) in a review of these studies showed that on average species richness was 30% higher on organic farms than conventional. However, they also found that in 16% of studies organic farming was associated with a decline in species richness. Overall, not only does species richness appear to benefit, but on average, organisms are 50% more abundant on organic farms. In another review of the subject Hole *et al.* (2005) found that birds, mammals, invertebrates and arable plants all benefited from organic management. Species that appear to do less well under organic conditions are non-predatory insects and pests. These reported shifts in species occurrence and abundance associated with organic agriculture may represent the development of new equilibrium communities. This begs the question raised above, are some species/communities more desirable than others? Against the background of dramatic declines in diversity in agricultural ecosystems, arguably anything that reverses this trend should be regarded as being desirable.

Many of the studies comparing the impacts of organic and conventional agriculture on biodiversity can be criticised on methodological grounds, because it is very difficult to find replicated comparable farms and because of time and scale effects discussed in Chapter 9. This may explain Bengtsson *et al.*'s (2005) observation that the effects of organic farming are most marked in studies carried out at the scale of plots rather than farms. Timescale and land-scape effects are likely to be more problematic in assessing the impacts of organic agriculture in the uplands. In addition to this problem the differences between organic and conventional farming tend to be less marked in the uplands. These factors may have contributed to the lack of comparative studies based in the uplands.

The impacts of organic agriculture on biodiversity have nearly always been assessed in comparison with conventional agriculture and less frequently mon-itored as change over time following conversion. A problem with this comparative approach is that conventional agriculture is only loosely defined and encom-passes a wide range of different farming practices. At the moment we have little idea of the relative environmental impacts of organic farming compared with a conventional farming approach adhering to the prescriptions of an agri-environment scheme (Hole *et al.*, 2005). This lack of information makes it difficult for policy makers to decide how best to fund enhancements in biodiversity.

The language of low-impact farming and conflicts with conventional agriculturalists

Although landscape modelling may offer a rational scientific method to decide what farming practices are best included or excluded from low-impact farms, the history and language of the environmentally friendly farming 'move-ment' appears less scientific. Phrases such as 'the organic movement' and 'converts to organic agriculture' give the impression of religion rather than science, so it is perhaps not surprising that tradition seems as important as science in determining 'what thou may and thou may not do'.

Organic agriculturalists regularly use phrases such as 'farming as nature intended' which have overtones of an Earth goddess. At first glance, these phrases have a ring of good old-fashioned values and common sense. However, from a scientific point of view the idea that farming in any shape or form could be described as 'natural' is nonsensical. Furthermore, it begs the question what is nature anyway? And if it can be defined – can it have intentions? Another phrase frequently used by the organic movement is 'farming in balance with nature', but as we have discussed above there is no one balance of nature. We need to decide what balance of food production and environmental conditions is acceptable, rather than devolve responsibility to nature, and this decision-making process may need to be better informed about exactly what levels of ecological service provision are associated with different farming systems.

The apparently quasi-religious language used by some elements of the low-impact farming movement is indicative of a very different world view from that held by many traditional scientists. Without wishing to partition blame, such different world views have contributed to a lack of understanding, collaboration and even trust between some scientists involved in agricultural research and some practitioners of low-impact farming. Another factor contributing to this suspicion is that the aims of conventional agriculture (although only rarely explicitly stated) primarily revolve around the levels of food production required to sustain an expanding human population. While this is not necessarily incompatible with the aims of low-impact farming, it may have resulted in the two parties arguing passionately about different objectives, all of which are laudable. This mistrust has been manifested by scientists working in the field of organic agriculture avoiding publishing in established agricultural journals (Watson et al., 2006). This is regrettable for all concerned.

Summary

This chapter has reviewed the main agricultural systems that have been designed at least in part with the aim of producing environmental impacts that are considered more desirable than those associated with conventional agriculture. A range of different indexes of environmental change are available with which to measure the success or otherwise of low-impact farming. However, because environmental change is so multifactorial there is no single agreed standard index. Perhaps because of this the environmental objectives of low-impact farming systems are typically imprecisely defined. In contrast the rules which govern what management practices are allowed within different low-impact farming systems are generally very tightly defined and regulated. While this may be good for ensuring consumer trust, and tightly defined rules are easy to police, the lack of measurable environmental objectives means that approved management prescriptions may be followed, but little environmental benefit may accrue. If low-impact farming systems are to be more successful in enhancing the environment, they need to be scientifically monitored and assessed against an agreed index. Furthermore approved management prescriptions should be designed and chosen on the basis of ecological understanding, scientific field trials and landscape modelling rather than on the meaningless notion of what nature intended.

9

Landscape and farmscape ecology

Introduction

How big are the average farm and the average field? Well obviously that depends on the type of farm, which in turn is related to its location. Within Europe, low-productivity upland farms have historically covered vast expanses of open, often communally grazed hill and mountain, while on the more fertile soils of the lowlands smaller mixed farming enterprises have been based on fields enclosed by hedges and walls. In contrast, in North America the larger farms are often associated with the flat mostly arable prairies. Farm sizes not only vary with geography but have also changed over time with historic human events; European colonisation, feudalism and land-ownership legislation have often been important factors. These have often worked in combination, because feudal overlords were attracted by the more fertile lowland areas, which were able to support more of a peasantry based on large manorial farms and small tenanted plots. In addition, cultural traditions of inheritance such as dividing the family farm equally between all siblings, or alternatively all the land passing to the oldest male child, have also influenced farm sizes. These factors have resulted in striking regional variation in farm size, for example in the United States (see Figure 9.1) and also in Europe where the average farm size is only 14 ha, while in the United Kingdom as a whole it is 77 ha and in Scotland 123 ha. In the modern period, intensification and mechanisation have tended to increase average farm and field sizes, while interest in hobby farming and the organic movement has moved in the opposite direction. Similar pressures have also influenced field shapes as well as field sizes; small irregular traditional field patterns were frequently defined by constraints of topography, the amount of land that could be practically worked, the availability of stone, the cultural norms or current agricultural practice.

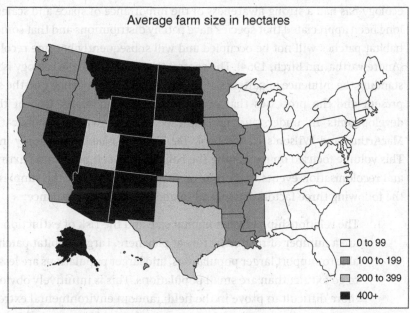

Figure 9.1 Variation in average farm size within each state across the United States. Data derived from US National Agricultural Service Statistics.

The point of the above introduction of the vast number of factors that have influenced the sizes and shapes of the management units of agricultural production (fields and farms) is that very few of these factors have any ecological significance. Of the species that inhabit agricultural land none of them recognises either the farm-scale or the field-scale. This may seem obvious with flocks of birds that migrate over large distances to feed, breed and roost, but surprisingly many invertebrates are also highly mobile (Topping, 1999). Similarly many ecological processes, such as the movement of water and nutrients, occur at scales considerable larger than the average farm. At the other end of the spectrum, certain agriculturally important ecological interactions are very fine-scale (Urban, 2005). Over recent decades this disparity of scale between ecological processes and agricultural management has become apparent. This chapter focuses on this developing understanding, and how it may be applied in practice to improve the environment, enhance populations of beneficial species and control populations of pest species.

What is landscape ecology?

Ecology as a science is about understanding why species occur where they do and why they are absent from other areas. Thus, from its foundation, ecology has had a strong awareness of the importance of space and scale; it has long been appreciated that species have patchy distributions and that sometimes habitat patches will not be occupied and will subsequently become recolonised (Andrewartha and Birch, 1954). The central theme of landscape ecology is under-standing the influence of patterns of environmental heterogeneity on the species present and the processes that occur within the landscape. One of the key developments in understanding landscape ecology was the publication of MacArthur and Wilson's classic book *The Theory of Island Biogeography* in 1967. This volume focused on explaining the balance between population extinctions and recolonisation events that occur on maritime islands. At the simplest level the following three factors are central to understanding this balance:

1.   The relationship between habitat size and the risk of extinction. There are a number of related factors at play here. Larger habitat patches are able to support larger populations, and larger populations are less likely to go extinct than are small populations. This is intuitively obvious but more difficult to prove in the field; random environmental extremes and inbreeding may both be important factors in increasing the chance of extinction in small populations. MacArthur and Wilson also described the related species–area relationship, illustrated by the fact that larger maritime islands are inhabited by more species than are smaller islands. This relationship can be described by the equation $S = CA^z$ where S is the number of species, A is the area, and C and Z are constants, determined by the group of organisms and the degree of isolation of the island.

2.   The relationship between habitat isolation and recolonisation rate. This was demonstrated in a dramatic experiment, in the Florida Keys, that involved killing all the invertebrates on a series of small islands located at a range of distances from the mainland. MacArthur and Wilson found that the more isolated islands had lower rates of recolonisation. Remote islands/habitat patches tend to support fewer species than do islands adjacent to the mainland or less fragmented habits.

3.   The natural limits to a population's ability to increase in size. Different species have different birth-rates and hence different abilities to recover from crashes in their population. Thus over time, species with low birth-rates are more likely to be driven to extinction by repeated random cata-strophes than are species that are able to quickly recover between episodes.

The scientific synthesis of the above and additional elements into the current view of landscape ecology is highly complex; however, in the context of the agri-environment, metapopulation theory can be simplified as follows.

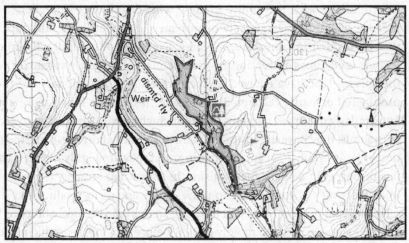

Figure 9.2 Many agricultural landscapes contain islands of semi-natural habitats such as woodlands. The populations found in small isolated patches of habitat are more likely to become locally extinct and less likely to re-establish than are populations found in large blocks of habitat.

Remnant patches of semi-natural habitats (woods, wetland, species-rich grasslands, etc.) can be considered as islands in an uninhabitable sea of intensively farmed land. Examples of isolated patchy woodland habitats can be seen in Figure 9.2. Populations that inhabit small and isolated habitat fragments are more likely to go extinct and are subsequently less likely to re-establish than are those in large, non-isolated areas. Good quality habitat patches that are net exporters of individuals are called 'source' populations (birth-rate exceeds deaths), whereas degraded habitats are likely to be net importers of individuals and are termed 'sink' populations (death-rate exceeds births). The relative significance of local extinctions, movement and re-establishment are not just landscape dependent but will vary between species. Those species with highly specific habitat requirements are more likely to live in isolated locations than are generalist species, which are able to survive if not prosper in a range of habitats. Species that are highly mobile will have higher recolonisation rates than do more sedentary species. Species that have low birth-rates will be prone to extinction in an agricultural landscape where farming activity regularly disturbs areas by ploughing, cutting, grazing, etc. and in doing so kills many of its slower moving inhabitants. The ability of the population to rapidly recover following perturbation by agricultural activity is a vital survival aid in the dynamic farmed landscape.

Although change is a universal constant, some habitats are more dynamic than others. Evolution has equipped the species that inhabit rapidly changing environments with the ability to cope with disturbance and change. In the absence of humans, large-scale disturbance events are generally rare, as are the species that inhabit disturbed sites. Examples of naturally disturbed sites include sand-dunes, retreating ice-sheets and recently active volcanoes. Many of the species that have evolved in these rare disturbed habitats are therefore preadapted or partly readapted to life within the now much more abundant but similarly disturbed agri-environment. Such species are known to ecologists as ruderals; they have life-histories characterised by short lifespans, the production of large numbers of small offspring, and they are highly mobile and invest few resources in defence mechanisms. In short their strategy to cope with life in a changing environment is to move about and/or to produce lots of small offspring that move around and by doing so increase the probability of finding the next suitable patch to rapidly exploit before moving, reproducing or dying. Confusingly these species are also referred to by ecologists as 'r' strategist species, where 'r' does not refer to ruderal, but to a term used in simple population models to represent the finite rate of increase (the number of offspring an individual can produce under ideal conditions). This maximum potential to reproduce r is key to understanding the population biology of ruderal species, which are prone to dramatic crashes and increases in the size of their populations. This is the life-history strategy of the typical agricultural weed or pest species; such species inhabit agricultural land. At the other end of the spectrum slower growing, less mobile species, which produce few offspring, are the types of species that now find themselves in small isolated populations in patches of suitable habitat in a sea of hostile agricultural land. Their continued survival is dependent on enough patches of habitat remaining inhabited to enable re-colonisation to occur following any local extinction event. These species are referred to by ecologists as 'K' strategists, where 'K' denotes the carrying capacity (the stable equilibrium population that the species reaches when it is not disturbed after many years).

More complex landscapes and more complex species

Species that adhere to life-history caricatures such as 'r' rapidly reproducing ephemeral species or 'K' slow-growing, long-lived sedentary species interact very differently with their landscape (see Chapter 1).

1. 'r' strategist species, being highly mobile, are likely to benefit from some habitat disturbance and the long-term viability of their populations is likely to be enhanced by landscapes containing mosaics of regularly disturbed habitats.

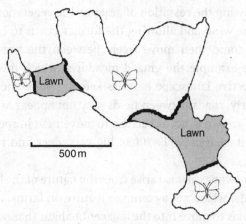

Figure 9.3 When E. B. Ford visited the island of Tean in the Scilly Isles before 1950 three populations of meadow-brown butterflies were separated by two small grazed lawns which were sufficiently exposed to discourage butterflies from crossing the island. Once grazing stopped the lawn vegetation became tall enough to provide shelter and allowed the butterfly populations to mix.

2.   'K' strategist species, being slower growing and reproducing in older age, are likely to be negatively affected by disturbance, and the long-term viability of their populations is likely to be enhanced by landscapes containing large blocks of undisturbed habitats linked by networks of corridors.

In reality of course things are rarely this simple. Landscapes, particularly farmed landscapes, are complex mosaics that frequently change over time. Mountain ranges, open water and woodland blocks may act as barriers to movement (Vandyke *et al.*, 2004), whereas rivers and other linear features may act as natural wildlife corridors. The underlying geology, human activity, the occurrence of invasive species and many other factors may all influence habitat quality and/or patch size.

In practice the landscape barriers preventing or reducing the movement of individuals between populations can sometimes be surprisingly open and small. In a study of the distribution of meadow-brown butterflies (*Maniola jurtina*) with different numbers of spots on their wings on the island of Tean in the Scilly Isles, Ford (1964) reported a significant mixing of populations between 1953 and 1954. The increased movement of butterflies between previously isolated populations on higher ground was attributed to the removal of cattle from the island in 1950. While the island was being grazed it was broken into discrete habitat patches by low-lying lawns (see Figure 9.3). Butterflies were observed to find these areas inhospitable and would not fly across them because of the risk of being blown out to sea. Following the cessation of grazing, the vegetation grew tall, offering shelter from the wind and allowing the former lawns to be colonised by butterflies which could then move freely between the previously isolated populations. In this example the grazed meadows, which measured about 500 m wide, were effective landscape barriers keeping populations isolated for many years. Similarly, roads between fields may not appear as major obstacles to humans, but are known to act as barriers to movement in species as diverse as bumblebees (Bhattacharya *et al.*, 2003), big cats, deer and reptiles (Ng *et al.*, 2004).

Complexity within the landscape can also arise from the nature of the habitat edge. Abrupt edges to habitats, which are a common feature on farms, may be less likely to encourage animals to move into the adjacent habitat than would a soft fringe where one habitat gradually merges with the next. Measures of habitat isolation that take into account factors such as this and are based on area informed measures, for example the amount of suitable habitat within a given radius (rather than nearest-neighbour distance), are better predictors of immigration rates than the earlier simpler measures (Bender *et al.*, 2003). Species adapted to life within large unbroken blocks of habitat are known to thrive less well close to the edge of their habitat (Huggard, 2003). The decline in habitat quality around the fringes of habitats or the edge effect is often related to increased rates of predation. For such species habitat fragmentation, which increases the proportion of edge to core habitat, is therefore associated with a marked decline in habitat quality. Similarly, habitat patches with irregular shapes, i.e. with a high proportion of edge habitat, are likely to be poor habitat for these species. In contrast species associated with habitats with a long history of human disturbance, such as *Boloria euphrosyne* (the pearl-bordered fritillary) in coppiced woodlands often thrive better in fringe habitats, and their management is typically associated with increasing the amount of available edge habitat such as creating woodland glades and rides (GreatorexDavies *et al.*, 1992).

Landscape ecologies can differ dramatically between species with similar dispersal mechanisms, even within the same landscape. In a study of lichens on isolated habitat patches (gravestones) Warren (2003) found evidence of isolation by distance (isolated gravestones had lower rates of colonisation) in *Candelariella vitellina*, which was generally confined to the rarer sandstone gravestones. In contrast no isolation by distance was observed in the same graveyard with the lichen *Placynthium nigrum*, which was associated with the more abundant granite gravestone habitat. Furthermore the isolation by distance observed in *C. vitellina* was landscape and time-period specific. The relationship was not found in larger graveyards where the few *C. vitellina* colonies on the granite headstones were sufficient to effectively stop the few sandstone graves from being isolated. In addition the low rates of colonisation of isolated headstones was only found within the first 50 years after their erection; subsequently spread from within the stone masked the effect.

Complexity in how individuals interact with their landscape is also known to occur within species. Resident populations of the North American waterbird killdeer (*Charadrius vociferus*) are known to be more sedentary than are winter residents or transient birds (Sanzenbacher and Haig, 2002). In butterflies, males have been found to move more rapidly than females across agricultural land-scapes (Haddad, 1999) and starving predatory beetles increase their rate of movement through hedges (Mauremooto *et al.*, 1995).

The ecological complications outlined above demonstrate that if we are going to practically apply landscape ecology theory within the farmed landscape then in many cases the impacts of landscape-scale management are likely to be species and landscape specific. For example we have seen that changes in grazing practice can result in populations merging or conversely being isolated, while other species would be unlikely to be affected by such changes. The simple metapopulation idea of island hopping has been replaced by a more complex understanding of species-specific landscape-scale spatial dynamics.

Landscape ecology and the agri-environment

We know that species interact with the landscape which they inhabit and that this may be important in determining the long-term fate of the species. But why does this matter to the agro-ecologist? The farmed landscape is the most dynamic and managed of all landscapes and agriculture is about managing populations of different species, encouraging some species and discouraging others. Typical agricultural landscapes are constructed of small patches of iso-lated semi-natural habitats and as we have seen both small areas and fragmen-ted habitats can be associated with declines in populations. Therefore understanding landscape ecology is a vital tool for the modern agriculturalist and conservationist.

Landscape ecology and metapopulation theory in particular has been in vogue for about 20 years. The oversimplified concept of isolated populations inhabiting islands of semi-natural habitat in a barren sea of agricultural land is one that has greatly influenced the thinking of conservationists generally and particularly in an agricultural context. Out of this view of the agri-environment has come an emphasis on encouraging the movement of wildlife across agricul-tural land to reduce the effects of population isolation and enhance population persistence. Farm conservation planning has been dominated by talk of wildlife corridors, linear features enhancing connectivity, conservation headlands, buffer strips and shelter belts. Habitat creation projects have been designed to create stepping-stone habitats to link fragmented populations. But is this obsession built on good ecology or on the practicality of abandoning farmed land to intensification and focusing on what is salvageable in the boundary habitats?

Boundary habitats, linear features and wildlife corridors

Field-margin boundary habitats include a range of linear habitats that have been given high conservation status within the farmed landscape. These include dry-stone walls, hedges, shelter belts, fences, ditches, abandoned roads, streams and even grass-margins. While they are considered as sanctuary habitats within the agricultural landscape their importance as wildlife corridors has been stressed. The significance of boundary features as habitats in their own right has not been overlooked particularly when considered along with the striking declines in lengths of hedges etc. that occurred when field-sizes increased during agricultural intensification (see Chapter 3). Hedges in particular are considered to be of high conservation value because of the wide range of species they support. They have been the basis of many ecological studies which demonstrate that botanical and structural diversity within hedges is reflected in higher numbers of species utilising the resource.

From a conservation standpoint, the management of boundary habitats has been designed to enhance their habitat quality by encouraging both botanical and structural diversity. This also results in the agricultural benefits of encouraging generalist predators and pollinating insects and suppressing weed species. Boundary habitats have been promoted by conservationists to the agricultural community as:

1. windbreaks, protecting crops, sheltering livestock and preventing soil erosion;
2. effective field boundaries;
3. habitats for natural predators and pollinators;
4. habitats for game birds;
5. a way of attracting grant funding;
6. a way of improving public perception of the industry;
7. buffer zones, reducing the drift of agrochemicals;
8. a way of controlling rabbits.

The more sceptical members of the agricultural community would argue that these benefits have been overstated by conservationists and organisations interested in promoting rural sports. They would claim that boundary habitats encourage crop pests and weeds and are a waste of potentially productive land. In fact both these points of view have merit. Well-managed grass-margins, which are not disturbed, spread with fertiliser or sprayed with pesticides, have been demonstrated to suppress weeds and encourage beneficial insects, and thus reduce the requirement to spray. In contrast, field-margins that are regularly disturbed either physically or by herbicide applications and have elevated soil fertility are likely to be dominated by undesirable weedy species with the potential to spread.

Similarly, sceptical ecologists might claim that the case for boundary habitats enhancing conservation value by encouraging movement between isolated populations is not entirely convincing. In particular there is evidence that hedges and other boundary features may actually impede the dispersal of flying insects and ground-moving beetles. This is easy to imagine with weak-flying woodland butterflies such as the wood white (*Leptidea sinapis*), which may not readily fly over hedges or cross gusty gaps in hedgerows. However, the movement of stronger flying species such as hoverflies is also known to be impeded by hedges (Wratten *et al.*, 2003). The weight of evidence seems to support the general principle of landscape ecology that linear features do indeed encourage the movement of wildlife between fragmented habitats, but this is not always the case. Other criticisms that have been directed at wildlife corridors include:

1. The possibility that they may increase the risk of predation as carnivores may target individuals moving along linear features.
2. They may bleed good habitats of wildlife by encouraging animals to move into less suitable habitats.
3. They may encourage breeding between previously isolated and locally adapted populations, resulting in new maladapted hybrid genotypes, a phenomenon known as outbreeding depression.

There is little evidence to support these arguments, but they remain viable possibilities worthy of further research. A significant problem facing researchers trying to estimate the extent that linear features encourage the movement of wildlife is that the effects of habitat fragmentation are most significant with sedentary species with exacting habitat requirements, and therefore by definition they are trying to quantify rare movement events. Looking at this problem another way, the majority of species that inhabit old agricultural landscapes such as those of Europe tend to be highly mobile generalist species and although they may well readily move along linear habitats, they are less likely to be detrimentally affected by habitat fragmentation. Thus the significance of linear features as wildlife corridors might be greater in landscapes such as the Americas and Australia that have been more recently fragmented by agricultural activity and which contain more species that only thrive in undisturbed habitats.

After considering the above complexities it is clear that many new hedges and other linear habitats are established within the agricultural landscape without a full assessment of the possible ecological implications being made. For example, a hedge established with the apparently sound ecological aim of connecting two isolated blocks of woodland may encourage the movement of badgers (*Meles meles*), thus transmitting bovine TB between adjacent dairy farms, while simultaneously increasing the isolation of habitats containing populations of butterflies.

Managing time and space using ecological models

Species that live within the agricultural landscape carry out their lives in a diverse and dynamic environment; they may reproduce, survive the winter and find food etc. in some patches better than others. Thus if we wish to understand how populations of either desirable or undesirable species interact with the farmed environment (with the ultimate aim of managing these populations) then firstly we need to know exactly how the environment changes over time and space, and secondly we need to understand how this affects our target species. These two aspects are key elements in the construction of landscape models, which have become important tools used in predicting the likely ecological impacts of possible changes in future land use.

Fortunately, describing how farmed landscapes change over space and time has been greatly assisted by administrative systems linked to paying subsidies to farmers, which are related to the extent of land under different managements. Typically, where these systems are in place, field locations are identified by codes and linked in databases to details of their area, and agricultural use in any one year. Even in regions where records are not precisely defined on a field-by-field basis, data on regional average land-uses allow different cropping systems and pastures to be ascribed to fields within Geographic Information Systems (GIS). At a finer temporal scale, local agricultural knowledge about when crops are typically sown, sprayed, harvested and the ground recultivated etc. allows a virtual landscape to be constructed with a computer that changes in seasonal or monthly increments. Thus within a computer we can assemble a picture of the farmed landscape which changes over time in a way that represents the real landscape in which our species live, reproduce, disperse and die. Such computer systems can be direct field-by-field copies of real landscapes or simplified model landscapes which encapsulate the main attributes of real landscapes (see Figure 9.4).

Once the farmed landscape has been adequately described, the next stage in creating a spatially and temporally explicit ecological model is to incorporate

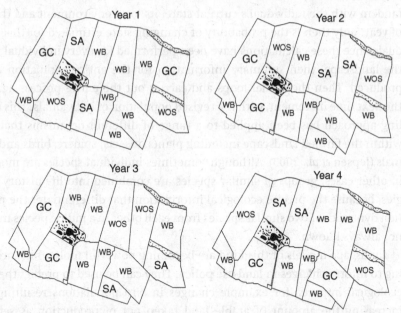

Figure 9.4 The basis of many landscape ecology models are GIS type databases which describe the farmed landscape. Temporally and spatially explicit data such as those illustrated above show how the locations of different crops within a farm change over time. GC = grass/clover pasture, SA = set-aside, WB = winter barley, WOS = winter oil-seed.

the associated ecological data. This may involve much time-consuming field-work in which estimates of survival, reproduction and dispersal rates in each of the different habitat patch types are generated. Alternatively the ecological data may include the flow of nutrients or water through different land types. Because such models simulate change over time, these ecological parameter estimates may be required for each time increment. Additional complications may be incorporated if there are significant interactions between habitat patches or if the size or specific location of patches affects the life-history parameters. In many cases these ecological parameter estimates have been extracted from the published literature.

Once the above two elements of a spatially explicit model are combined the model can be run. Within the model every individual of the defined species is discretely represented mathematically. The starting population sizes are gener-ally based on census data. At every specified time increment, each individual within the population is considered (at random) and how it interacts with its environment determined. It may change its state, by growing, reproducing, moving or dying, etc. These changes are determined by life-history rules defined by the ecological information compiled earlier. The rules are considered in tandem with the individual's current state, its local environment and the time of year; only then is the probability of changing state estimated for that individual. Once these calculations have been performed for every individual within the landscape, then summary information for the entire population can be produced. Then the model loops and carries out the entire process again for the next time increment, with the revised population of individuals. This modelling approach has been applied to a range of different organisms that occur within the farmed landscape including plants, insects, spiders, birds and mammals (Jepsen *et al.*, 2005). Although sometimes individual species are modelled, in other cases, groups of similar species are combined into life-history strategies, because the precise ecological information that differentiates the population dynamics of two similar species from each other in a multi-species model is not always known.

Landscape models of this kind are becoming powerful tools which are starting to inform changes in land-use policy. They can be used to predict the likely ecological impacts, for example changes in bird populations resulting from increasing the amount of arable land taken out of production as set-aside. More subtly the same model can be used to simulate the effect of encouraging farmers to locate land taken out of production in specific places within the landscape, such as along or around field-margins, or concentrating it in large blocks or along river edges. Perhaps more importantly the development of the landscape modelling approach has happened in parallel with a change in the way that we think about the farmed landscape. At the start of the twenty-first century the countryside is seen as a multifunctional landscape that does a lot more than produce food (see Chapter 2). The farmed landscape is now being considered more holistically as delivering many ecological services such as pure water, flood prevention, providing habitats for wildlife and offering many ways of supporting rural human communities including the production of food. More complex models can reflect this and simulate the effects of policy changes across a range of different outcomes (Warren and Topping, 1999).

Although landscape models are being utilised in informing proposed policy changes they are not yet widely used in directly informing policy or management regimes designed to produce a specific outcome. This is probably because of limitations of the real world. It is not often that an individual or organisation is in the position to manage at the landscape-scale over the long term with the goal of managing a single target species. The closest examples probably include the repeated coppicing of blocks of woodlands to produce a long-term supply of suitable habitat for butterflies. Arguably farmers come closest to fitting this description, and in this case the vagaries of the markets, climate and subsidy changes all complicate long-term, large-scale planning.

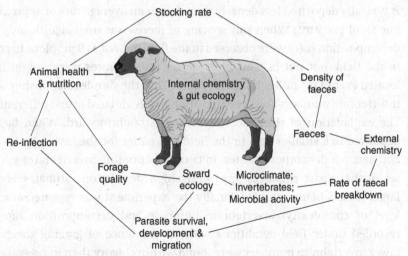

Figure 9.5 Novel forages such as chicory may reduce the numbers of intestinal parasites in grazing animals by accelerating the decomposition of faeces so that parasite larvae do not have enough time to develop in the field. However, this effect is scale-dependent with the opposite result being obtained in small plots where chicory is found to delay the decomposition of faeces.

The importance of scale in landscape ecology

Since the early days of experimental ecology it has been apparent that the results obtained in the laboratory or glasshouse can be very different from those obtained under field conditions. These differences between small experiments and field observations are often a result of scale. Many ecological processes work at the large scale and not the small (or vice versa). Therefore scale can be a real problem for the landscape ecologist. Just because we understand how an ecological process works within a field, or at a farm-scale, it does not necessarily mean we can extrapolate to predict what will happen at a larger scale.

An example of how scale can affect the outcome of field experiments in agro-ecology is demonstrated by the relationship between invertebrate population size and the rate of decomposition of faeces within different pastures (Williams and Warren, 2004) (see Figure 9.5). Novel forages such as *Cichorium intybus* (chicory) are known to reduce the numbers of intestinal parasites in livestock but the mechanism/s are unclear and may apply at any stage in the complex life-history of the parasites.

In small plots measuring 1 m^2, conventional ryegrass–clover pasture vegetation contains more invertebrates and faeces decompose more rapidly than in plots containing chicory. There is a simple relationship at work: more invertebrates result in faster rates of decomposition. In the field, however, sheep faeces is typically deposited less densely than this, at an average rate of approximately one stool per $9\,m^2$. When this spacing of faeces was used significantly slower decomposition rates were observed in the field but not in $9\,m^2$ plots. In contrast, in the field (but not in $9\,m^2$ plots) fewer invertebrates were caught in traps located every $1\,m^2$ rather than every $9\,m^2$. Thus the simple relationship between invertebrate numbers and decomposition rate is violated across different scales. The explanation of this effect of scale is straightforward. When faeces are deposited in a smaller area in the field it is easier for the invertebrates to find and hence it decomposes faster. In contrast sampling invertebrates with traps set further apart produces apparently high population estimates because a larger area is being sampled. Finally the experiment was repeated at a higher level of complexity: invertebrate numbers and decomposition rates were recorded under field conditions with the presence of grazing sheep. Again lower invertebrate numbers were found within chicory than in ryegrass–clover pasture, but this time, the faster rates of decomposition were observed in the chicory plots. This time the complexity of the larger-scale observation is not as simply explained and the difference probably relates to different grazing and defecation habits in the two pastures. When grazing chicory, sheep tend to produce a latrine area, and therefore there is much less faeces available within the rest of the sward, resulting in a higher invertebrate:faeces ratio in spite of the lower absolute numbers involved. Whatever the reasons are that cause these results to differ at different scales, they neatly illustrate the dangers of extrapolating from small experiments to the farm-scale.

The evaluation of field- and farm-scale trials

The complexities caused by ecological processes varying at different scales are probably most problematic for the agroecologist when evaluating the results of field- or farm-scale trials. This is partly responsible for the controversial nature of comparisons made between organic and conventional agriculture or assessments made concerning the likely ecological impacts of growing genetically modified crops. The problems of extrapolating results to different scales means that it is virtually impossible for ecologists to provide definitive answers to these sorts of questions.

Let us explore this problem further, but avoid these controversial issues and instead consider the evaluation of a field trial to assess the advantages and disadvantages of reducing agricultural inputs. Imagine a field trial that compares two similar cereal fields. The first uses reduced but targeted inputs of pesticides, herbicides and fertilisers, and it incorporates grass-margins around the field to encourage beneficial invertebrates and thus reduce the need for insecticides. In the second, control treatment, plot, non-selective pesticides and herbicides are used pre-emptively at higher rates, and there is no management of field-margin. At the field-scale the outcome of the trial is simple, the reduced input treatment produces significantly lower yields than the control treatment; however, it also has reduced costs, because of its lower inputs and reduced fuel use. Overall the net economic impact is identical for both treatment plots. So can we advise a farmer that reducing their use of agrochemicals will not affect the profitability of their business? The answer is – no, because of scale effects, it is possible that if the low-input treatment was applied to every field on the farm the results could be very different, and they could be better or worse.

It is possible that the full advantages of trying to encourage beneficial invertebrates are less than 100% effective in an isolated field. There may be little or no benefit of increasing the numbers of ladybirds (*Coccinellidae* family), money-spiders (*Cinyphiidae* family) and rove-beetles (*Staphylinidae* family), if they are killed by the pesticides that are applied in the adjacent field. In which case expanding the low-input treatment to cover an entire farm may avoid this problem and be more worthwhile than only having an isolated plot. Conversely, a single low-input field may be unfairly advantaged by the fact that the pests in all the surrounding fields are being effectively controlled by high-level pesticide use. In this scenario, if the low-input treatment was expanded it may result in greater pest problems and more yield reduction than was seen in the isolated plot. Furthermore, the build-up of pest populations could worsen over time, adding an additional level of complexity in the extrapolation process.

While the above is a hypothetical scenario, the estimated yield of multi-lines of cereals based on extrapolating from small plots provides a real example. Multi-lines are varieties of crops which actually consist of combinations of different strains of the same crop. The strains are selected to be ready for harvesting at the same time and at the same height. They are also chosen so that they contain different disease resistance genes, and hence the entire multi-line is less likely to be susceptible to a particular strain of pathogen. Multi-lines are more frequently grown in the United States than in Europe. In the field they typically yield between 10% and 15% more than is predicted by their performance in agronomic trials. The reason for this is that the plots used in field trials are not only smaller than real fields, they are more uniform. Even in the most extensive monoculture annual crop fields in the United States, there is environmental heterogeneity not found in the trial plot. There are patches where the soil is deeper, or drier, or on a slight slope etc., and this heterogeneity is more efficiently exploited by a multi-line where different strains will have slightly different optimal growth conditions. This advantage has no way of being expressed within a uniform field plot and so the yield estimates are lower.

Macro-ecology

In theory it is a relatively simple conceptual scaling-up exercise to develop the landscape-scale models considered above into systems able to predict the outcome of land management changes on a national or international scale. However, as we have already seen in ecology the scaling-up process can hold many surprises. As before the first set of problems typically surround the quality of the data upon which the required GIS are constructed. Input data can be derived from remote sensing, from satellite images and/or from aerial photography from manned or unmanned craft. High-resolution remotely captured images can identify details such as vegetation type at the scale of a metre or two. This facility in itself has great potential as a management tool as it allows the identification of small patches of vegetation that are at risk from contamination from adjacent genetically modified crops (Davenport *et al.*, 2000). It can be used to search for needles in haystacks, such as sites likely to support rare or problematic species. If only a few sites are identified, it is a relatively easy matter to check them on the ground. However, as the number of sites or the number of different land-classes increases it becomes increasingly difficult to ground verify the data. This problem is exacerbated when there is a limited supply of field workers with the necessary skills. Similar issues can arise when the data are derived via the subjective interpretation of remote images. Person-to-person variation is easily detected if different workers are assigned different regions, and easily hidden if they are allocated small patches randomly spread across the whole target region.

An alternative approach at constructing the required geographic database is based on spatial interpolation and numerical process modelling (Aspinall and Pearson, 2000). Here other established geographic datasets, such as underlying geology, soil type, altitude, distance to nearest conurbation, etc. are combined to predict likely land-use classes. If annual government agricultural statistics are available they can greatly enhance the quality of the data. This method may also include an element of subjectivity in collecting the baseline data or in verifying the predictions, but generally this approach reduces the human aspect in decision making. As a result the possibility of misinterpolation of data is reduced, but anomalies can still arise from the mathematical overemphasis of particular datasets and therefore ground verification is still important. Warning signs include patterns of underlying geology being very clearly visible in land-use classes. Overall remote sensing has the advantage in being easier (if currently still costly) to update with datasets with short shelf-lives.

Whatever the source or fidelity of the baseline data an increasing number of macro-scale landscape models are being developed which predict a range of outcomes, for example: the occurrence of rare plants (Boetsch *et al.*, 2003), the distribution of fire risk (Hargrove *et al.*, 2000) and conurbation spread (Syphard *et al.*, 2005). While a variety of different modelling approaches are being utilised they all face similar difficulties in testing the robustness of their predictions. It is effectively almost impossible to perform an experiment at the landscape-scale and scientifically of little value. A researcher cannot insist that region X converts to organic agriculture while region Y continues to use pesticides, and even if this was possible, the regions would not be comparable because they are likely to differ in a host of other ways, e.g. in climate, soil type, etc. Furthermore the ecological outcomes of such change are likely to be very slow and it may take years for a significant change to be observed. Even so, historical records may provide data against which landscape models can be verified and these provide a degree of confidence in their predictive powers.

Summary

In this chapter we have discussed the fact that very few ecology processes actually occur at the farm-scale. Individual animals, plants, nutrients, pesticides residues, water, etc. may all move across field and farm boundaries in ways that we are starting to be able to understand and predict. Species typically associated with agriculture tend to be highly mobile and less habitat-specific than are those found in the islands of semi-natural habitat within the more intensely farmed land. Computer models are being developed that utilise this understanding to predict the outcome of policy changes or new technologies on the ecology of the agricultural landscape. The challenge for the future is to reverse the process and develop models which can define the prescriptions that are needed to sustainably manage a multifunctional agricultural landscape which delivers an adequate supply of foods and ecological services from a diverse and viable countryside.

10

The future of agri-environmental systems

Introduction

Agriculture has historically been location specific. The variability of climate, soil, topography and culture across a landscape resulted in a diverse and locally distinctive agriculture. During the twentieth century the Green Revolution saw dramatic increases in agricultural productivity and a more science-based agriculture with greater control of production factors such as chemical fertilisers, pesticides and machinery. Accompanying and, in many cases, supporting these changes was an increase in state subsidisation of agriculture, both direct production support and indirect support through research and development. This led to a decrease in location-specific agriculture and to more of a centralised or 'blueprint' agriculture driven by production targets (Fresco, 2002). What will agriculture and rural landscapes look like in the future? The final chapter of this book will consider the future of agricultural systems. Will agriculture be transformed to systems that integrate environmental protection and food production across the landscape to provide society with both food and fibre commodities and a wide range of ecological goods and services? Alternatively, will a more location- or region-specific system emerge with certain areas of land, and other resources, allocated to intensive production of food and fibre commodities, while farmers/managers in other areas focus on the production of ecological goods and services? Ecological theory suggests that in future more biodiverse agricultural systems may be more sustainable and productive than the monocultures that currently dominate agricultural landscapes around the world. However, there are considerable practical problems to overcome. This chapter will examine some of the important trends in global agricultural systems. The emerging trends in agricultural production systems, changes in technology, the predicted future changes in agri-environmental policy, the global trading system and the relevant agricultural trade agreements, the emergence of multilateral environmental agreements and how they are beginning to shape agri-environmental policy will be highlighted.

The context for agri-environmental systems

Agriculture does not operate in a vacuum. Agricultural management and production decisions are influenced by a range of factors from global trade to the changing structure of rural communities. Therefore, in order to understand the possible future trends in agriculture and agri-environmental systems we need to understand the trends in those contextual factors that drive agricultural management decisions. The following discussion will highlight some of the more prevalent of these contextual factors and the trends that are emerging. It should be noted that each of these factors does not operate independently and in many cases agricultural management changes are influenced by a number of interacting factors.

Market trends

Management decisions made by farmers are strongly influenced by the economic signals and economic incentives that are apparent to them. As such, local and global demand for food and fibre is one of the fundamental drivers of change in agricultural systems. The prices received for agricultural commodities and the costs of production inputs, in particular, have a strong impact on the pattern of management across agricultural landscapes. With respect to output prices, it has been estimated that global food demand will grow, which will put upward pressure on food prices. While it is predicted that global population growth will slow, other factors such as urbanisation of populations and robust growth in per-capita income will become relatively more important drivers in strengthening food and agricultural demand and, importantly, changing the type of food commodities demanded. An important determinant of global food demand is the distribution of the population across the world. For example, population growth in developing countries is predicted to slow but it will remain above the growth in developed countries resulting in the share of world population in developing countries, currently at 75%, continuing to rise (OECD/FAO, 2005). Urbanisation and income growth, particularly in some developing countries in Asia and Latin America, will result in consumers moving away from staple foods such as cereals and tubers to demand a more diversified diet with increases in meat, fruit and vegetable consumption (Table 10.1). This will have an environmental impact since diets rich in meats require the

Table 10.1. *Recent and predicted per-capita consumption of selected food commodities*

	Consumption (kg/capita/year)		Consumption growth (%/year)	
	2002–4	2014	1995–2004	2005–14
World				
Wheat	81.6	82.4	−0.36	0.27
Coarse grains	56.4	59.9	1.07	0.33
Rice	69.9	69.4	0.08	0.02
Meat	31.2	34.5	3.38	0.88
Vegetable oils	13.1	16.4	3.62	1.73
OECD				
Wheat	107.3	112.4	0.52	0.58
Coarse grains	114.5	131.6	3.30	0.70
Rice	19.6	19.0	0.42	−0.17
Meat	64.5	69.9	7.03	0.73
Vegetable oils	21.6	26.1	1.62	1.56
Non-OECD				
Wheat	75.5	75.7	−0.37	0.22
Coarse grains	42.4	43.8	−0.40	0.32
Rice	82.0	80.7	0.10	−0.06
Meat	23.2	26.5	1.64	1.18
Vegetable oils	11.0	14.3	4.64	1.91

Source: OECD and FAO secretariats (OECD/FAO, 2005).

production of feed-grains and meals and as such actually demand more cereal than diets based on direct cereal consumption. Greater meat production is also associated with concentrated nutrients in the form of manures. Greater fruit and vegetable production changes the intensity of land use, requirements for irrigation water and the use of pesticides. Therefore, changes in food demand will have a strong influence on the environmental impact of agriculture.

Increases in food production will require the extensification and/or the intensification of food production. In many regions there is little to no productive land that has not already been converted to agricultural production and in some areas the stock of productive agricultural land is decreasing due to urban expansion and/or land degradation (e.g. desertification). Nonetheless, increasing demands for food will expand the area of land used for agriculture, particularly in the developing countries, revealed as a predicted decrease in non-domesticated land. For example, it is predicted that between 1990 and 2050 the area of non-domesticated land will decrease by approximately 25% and 20% in Africa and west Asia respectively (Figure 10.1). During that same period the area

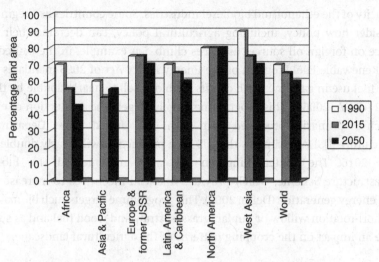

Figure 10.1 Non-domesticated land as percentage of total regional land area, 1990–2050. (Source: UNEP, 1997)

of non-domesticated land will decrease by less than 5% in Europe and the former USSR and remain unchanged in North America. The expansion of agriculture will come at an environmental cost as natural ecosystems such as rainforests, wetlands and native grasslands are cleared, drained and cultivated. Expansion of production in more arid landscapes will put pressure on ground and surface water resources to be used as irrigation water.

An important trend that may have a significant impact on agricultural production systems is the predicted decrease in global fossil fuel supplies, which will increase the price of fuel, fertiliser, pesticides and other production inputs as well as the cost of transportation of food. Higher input prices will have an impact on the viability of existing production systems and technology. It is likely that there will be a shift away from production systems that rely on the intensive use of these inputs. In addition, there will be an increased substitution of inputs that are less impacted by increased fossil fuel prices and the adoption of alternative technologies that are fossil fuel conserving. For example, greater quantities of legume-fixed organic nitrogen may be used as a substitute for higher priced synthetic nitrogen, or multi-cropping systems better able to utilise the available nutrient resources may become more prevalent.

Another impact of increasing fossil fuel prices is the increased importance of energy crops. The relatively high prices for fossil fuel will contribute to favourable comparative returns for the production of alternative fuels such as ethanol and biodiesel, which are often produced using grain or biomass from agricultural crops. This provides an economic incentive for expansion in the production capacity of the ethanol and biodiesel industries. Some countries are beginning to consider how policy, including agricultural policy, can decrease their dependence on foreign oil sources as prices climb. For example, in the United States the Renewable Fuel Program of the Energy Policy Act of 2005 mandates renewable fuel use in gasoline (with credits for biodiesel) to nearly double by the year 2012 (USDA, 2006c). This programme largely targets the production of ethanol, which is primarily produced from corn. The United States Department of Agriculture (USDA) estimates that US corn production will nearly double by the year 2016. The United Kingdom has also implemented the Bio-energy Infrastructure Scheme, which provides financial incentives to increase renewable energy generation (Defra, 2006). This programme targets such biomass crops as short-rotation willow or poplar, grasses, straw and wood fuel and as such will have an impact on the cropping patterns of the agricultural landscape.

Technological trends

Increases in food production, to meet the increased demand discussed earlier, will partly come from the expansion of agricultural production onto previously non-domesticated land. However, it has been estimated that increases in cultivated area will contribute less than 20% of the increase in global cereal production between 1993 and 2020 (Pinstrup-Andersen, 2001). Therefore, a more important route to increased food production, particularly in the developed countries in North America and Europe (Figure 10.1), will be from increased output per hectare. Technological change is strongly influenced by economic factors such as input and output prices, which are, in turn, influenced by factors such as policy and climate. It has been argued that this has been driven by a positive feedback loop, with technological advances in agriculture, producing more food from fewer people, allowing more industrial production, generating more wealth, some of which was put back into agricultural research. Changes in production technology include changes in machinery, computer technology (Global Positioning Systems (GPS) and precision farming) and crop and animal varieties, including genetically modified organisms (GMOs). In the next ten years it is predicted that technological developments that result in increased agricultural productivity, and therefore increase food supply, will offset the increased demand for food due to income and population growth (discussed earlier) such that commodity prices will not increase as much as could be expected by examining demand factors alone (OECD/FAO, 2005). These technological changes, if implemented in developing countries, may provide an environmental benefit by, for example, requiring less non-domesticated land to be converted to agricultural production or enabling management practices that are more resource-conserving. However, it is difficult to predict the nature of these technological changes, even in the near future, and whether the changes will have a net positive or negative impact on the agri-environment.

As discussed above, the nature of future technological change is impossible to predict. However, GMOs and precision farming represent two recent technological developments that deserve special mention here as they appear to have the potential to facilitate fundamental changes to agricultural systems. Genetically modified organisms are organisms whose genetic material has been altered using recombinant DNA technology that enables the combination of DNA from different sources (e.g. two different plant species). At this point GMOs developed for agriculture are primarily plants (e.g. oil-seed rape/canola, maize, soybeans) that are tolerant to specific herbicides or that have insect resistance (e.g. Bt soybeans). Future applications of GMOs may include plants that produce human vaccines against infectious diseases, plants and animals that mature more quickly, plants with greater drought tolerance, etc. The environmental impact of GMOs is currently the subject of great debate. It has been argued that the widespread use of GM crops will impact wildlife habitat. For example, in a study of the impact of growing herbicide-tolerant GM crops (GMHT) in the United Kingdom it was found that compared with conventional crops the GMHT fields tended to have fewer weeds and/or a different collection of weeds and therefore had fewer invertebrates such as butterflies and bees (Defra, 2005). The study concluded that if GMHT crops were adopted across the United Kingdom it could have a negative impact on a range of wildlife species. However, these differences in habitat were attributed to the changed management that the GMHT crops facilitate rather than being due to a direct effect of the GMOs. Other research has indicated that appropriate management of GMHT crops can provide good habitat for a range of species as well as decreasing overall herbicide use. Other concerns include the impact of broad-spectrum herbicides or plant insecticides on soil biodiversity, the creation of herbicide-tolerant 'superweeds' through gene flow from crop to weed species, animal welfare issues associated with GM high-yielding animals, as well as the broader landscape impacts such as crop monocultures encouraged by GM crops. The net impact of GMOs on the agri-environment may be significant but is unknown. It is also worthwhile to note that the potential uptake of GM crops in Europe may be limited by consumer concerns and a reluctance to purchase GM products while adoption has been quite extensive in other regions (e.g. the United States, Canada, Brazil).

Precision farming, or site-specific farming, uses remote sensing information technologies such as GPS and Geographic Information System databases (GIS) to make appropriate farm management decisions, primarily for crop production. Applications for precision farming technology include the recording of information on yield, grain moisture content, harvest rate, soil fertility, weed density, etc. and linking it to a GPS system so the information becomes site specific. The primary objective of precision farming technology is to enable the farmer to adjust rates of seed, fertiliser and pesticide applications based on the site-specific information. As such, this technology can assist farmers in decreasing input costs, increasing yields and decreasing risk. With respect to the agri-environment, precision farming can either increase or decrease the quantity of pesticide and fertiliser inputs. In general, it has been argued that the technology will enable fertiliser and pesticide application patterns that decrease the quantity of nutrients and chemicals that are transported offsite thereby decreasing environmental pollution. However, the technology may also make it profitable to farm land that was previously unprofitable (e.g. draining of more permanent wetlands) thereby imposing the environmental costs of agricultural extensification. This extensification is also encouraged by the potentially high fixed costs of the technology. Another application of the remote sensing technology that is worth considering is its use in monitoring agri-environmental indicators and management encouraged by agri-environmental policy. In these applications remote sensing may improve the efficiency of agri-environmental policy spending by enabling more specific targeting and monitoring of compliance.

Climate trends

Agriculture, with its dependence on weather patterns, is believed to potentially be one of the industries most affected by climate change (Adams et al., 1999). The main direct effects of climate change on agriculture will be through changes in factors such as temperature, precipitation, length of growing season and timing of seeding, flowering and harvest of crops as well as the potentially beneficial effects of increased atmospheric carbon dioxide concentrations (IPCC, 1997). Indirect effects will include potentially detrimental changes in diseases, pests and weeds. The impact of these effects on agriculture and agro-ecosystems will include a required shift in planting and harvesting dates and probably result in a need to change crop varieties currently used in a particular area. Traditional patterns of water use may no longer be viable and in many cases water use efficiency will need to increase. The shifts in crop production, changes in water regimes and potential expansion of irrigation may negatively impact the agri-environment by decreasing water quantity and quality, degrading or destroying wetlands, soil and wildlife and wildlife habitat. In addition, the introduction or strengthening of invader and weed species may negatively impact wildlife habitat, altering biodiversity and forcing the adoption of more intensive weed management. Perhaps most important of all, there is general agreement that in addition to changing climate, there is likely to be increased variability in weather, which might mean more frequent extreme events such as heat waves, droughts and floods. However, the impact of climate change on agriculture will probably be highly variable across regions, with some areas experiencing increased yields and an increase in the number of viable crop varieties while in other areas annual cropping systems may no longer be viable. In one extensive modelling study it was predicted that climate change and climate variability may result in irreparable damage to arable land and water resources in some regions with serious local consequences for food production, and these losses are expected to be felt most profoundly in developing countries with low capacity to cope and adapt (Fischer *et al.*, 2005). With respect to climate change adaptation, given the dependence of the world's food supply on a few crop varieties and the predicted increased variability in climate parameters, it has been advocated that agricultural systems must be made more resilient by ensuring greater diversity of crops, animals and management systems.

Climate change could have a significant impact on agricultural input and output markets. As discussed earlier, input and output prices are a significant driver of agricultural management decisions. The changes in patterns of management, crop varieties produced and productivity associated with climate change will impact markets and management in many areas. Further, regulations aimed at decreasing greenhouse-gas emissions could increase the cost of or restrict the use of fossil fuels thereby increasing the cost of production inputs such as fuel and fertilisers. The cost of transportation of food may also increase with regulations aimed at mitigating climate change. This will reduce the amount of air travel and may decrease the possibilities for international trade in, particularly bulk, agricultural products. One impact of this is that countries may attempt to meet the demand for products that were formerly economical to import by increasing domestic production. These, and other, changes in input and output markets due to climate change and climate change regulation may have a significant impact on the agri-environment.

Another factor related to climate change that may influence the future of agri-environmental systems is the role of agricultural soils as carbon sinks. With appropriate management, such as reduced annual tillage and increased use of perennial crops including trees and shrubs, the stocks of carbon stored in agricultural soils (and above-ground in the case of trees and shrubs) can be increased thereby decreasing the stock of carbon dioxide in the atmosphere. These carbon sinks are recognised within the Kyoto protocol as an accepted mechanism for countries to meet their greenhouse-gas emission reduction commitments (IPCC, 2000). In recent years, governments, even within countries that have not ratified the Kyoto protocol, have begun to implement policies and institutions to encourage the development of and increase carbon sinks. This could have a positive impact (e.g. increasing soil conservation management) or a negative impact (e.g. significantly altering existing prairie habitat with increased area of trees) on the agri-environment.

Human and social capital trends

Social capital comprises the cohesiveness of people in their societies including trust, reciprocity, cooperation and rules (Flora, 1995; Dobbs and Pretty, 2001). Social capital has been shown to be important to facilitate coordination and cooperation for mutual benefit, in essence lowering the cost of working together. Human capital is the skill, knowledge, health and fitness characteristics of the people within a farm, community or region. Human capital can be enhanced through access to formal and informal education and medical services. In many countries changes in the structure of farms and farm communities have had an impact on the social and human capital of agri-environmental systems. For example, characteristics such as the age of the farmer, the type of tenancy (rent, own), the size of the farm, the type of farm (e.g. specialised crop or livestock farms or mixed farms), the availability of off-farm employment, the viability of small towns and villages and the existence or absence of social networks in rural areas are altering social and human capital. For example, in Canada between 1990 and 2000 the number of non-business focused farms (e.g. 'lifestyle farms', hobby farms) increased by almost 40%. Currently the number of non-business focused farms in Canada is greater than the number of business focused farms with off-farm income becoming increasingly important (Figure 10.2). The change in farm structure and farm type has altered the social capital within agricultural landscapes and in rural communities. This loss in social and human capital will impact the capacity and willingness of farmers to address environmental concerns. Individually, farmers may not have the expertise and experience to successfully adopt environmental management prescriptions. At the community level it may be difficult for farmers to work together to address more regional agri-environmental problems. For example, the improvement of water quality in a stream may require the cooperation of a number of farmers adopting specific management prescriptions. However, without the appropriate management skills (human capital) and ability to communicate and cooperate (social capital) the environmental improvement will be difficult to achieve. Therefore, the quantity and quality of human and social capital will be an important factor in determining the future of agri-environmental systems.

Policy trends

The important role that policy plays in influencing and determining the nature of agri-environmental systems was discussed at length in Chapters 2 and 4 of this book. The agri-environment is impacted not only by policy aimed

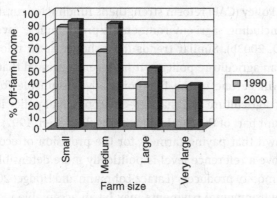

Figure 10.2 Percentage of total farm income from off-farm sources for small to very large farms in Canada in 1990 and 2003. (Source: Statistics Canada, 2005)

directly at agricultural management but also by more general policies such as property taxes, land zoning, urban expansion, food safety and hygiene and water policy. The nature of policy and policy trends responds to changes and trends in the preferences of society. As such, future policy trends will be directed by many of the issues discussed above including demand for food and fibre, input costs, climate change and the preferences and demands of the larger society. An alternative model has been implemented in New Zealand where, since 1986, the government has removed all policies that alter production or trade patterns. As discussed in Chapter 2 the level of producer support in New Zealand is currently the lowest of all Organisation for Economic Co-operation and Development (OECD) countries. Therefore, land management decisions and agricultural production are driven almost exclusively by market signals while potentially enabling the government to allocate expenditures and develop legislation that more directly addresses environmental concerns. In some cases this may be a preferable policy trend to a more indirect agri-environmental programme. However, it is impossible to predict future policy trends. The following represent some trends that have emerged recently that will impact agri-environmental systems.

The future of agri-environmental policy

Recent trends seem to indicate that agri-environmental policy will remain as an important force in many developed countries. For example, in the United States the '2002 Farm Bill' provided for an 80% increase in funding for agri-environmental objectives over 6 years while the European Union's 2003

Common Agricultural Policy (CAP) reform strengthens funding for rural development measures, including agri-environmental programmes, over the 2006–12 period (OECD, 2003b). Similar trends have shown up in reforms to Australian and Canadian agriculture policies. It has been suggested that environmental payments will be one of the few politically sustainable forms of government support to agriculture and that agri-environmental policy is set to become a more dominant part of the rural policy scene (Buckwell, 1997; Potter, 1998). It has been shown that paying farmers for the provision of ecological goods and services above a reference level is politically more defensible than paying farmers as commodity producers (Latacz-Lohmann and Hodge, 2003). In addition, these agri-environmental payments may be an acceptable means of agricultural support under World Trade Organization (WTO) rules, which will be discussed in more detail below. However, the structure that these agri-environmental policies will take is uncertain and the overall impact that the policy will have on agri-environmental systems is unknown.

In Chapter 4 it was discussed that an efficient allocation of resources is one that maximises social net benefits. Theoretically, the only way to determine the efficient allocation is to quantify all of the costs and benefits imposed by the allocation. An important step is understanding how much society is willing to pay for the ecological goods and services that are provided by the agri-environmental policy. For example, if an agri-environmental policy results in society paying, through programme incentives, more (less) for each hectare of wetland conserved than the value of the wetland to society then too much (too little) wetland is conserved. A rapidly growing area of research is focused on attempting to quantify the value of environmental public goods, including those that are provided through agri-environmental policy. Since many of these goods and services are not traded in markets (e.g. biodiversity, hydrological services) the values are derived using a range of survey and statistical techniques. While these valuation approaches are considered to be the preferred means to quantify the value of ecological goods and services, there are concerns about the validity of some of the estimates, particularly across landscapes and regions. However, without these values, determining an appropriate allocation of resources will be difficult in the absence of alternative approaches.

While agri-environmental policy expenditures are increasing it is important to consider the role of these policies in context of broader agricultural policy. Within the OECD countries support to agriculture from government payments represented over 30% of the gross farm receipts in 2004 (OECD/FAO, 2005). While recent changes in the composition of agricultural support have decreased the emphasis on production and trade distorting measures, market price support continues to dominate with agri-environmental programmes still being relatively minor components. For example, European agri-environmental policy expenditures represent less than 10% of the total CAP budget (OECD, 2003b) while projected conservation programme expenditures from the US 2002 Farm Bill are slightly greater than 20% of total Farm Bill expenditures (USDA, 2002). Therefore, the future of agri-environmental systems, while influenced by agri-environmental policy, will be largely dependent on broader agricultural policy developed to meet primarily non-environmental objectives. It should also be considered that social preferences could shift dramatically in the future such that it may not be deemed acceptable, for many reasons, for agri-environmental policies to be long-term initiatives. For example, climate change (as discussed earlier in this chapter) could result in local, if not global, food shortages and/or high levels of uncertainty about future production. In this case a socially optimal policy may involve increasing support for production to ensure adequate food, or, alternatively, ensuring agri-environmental health to increase the resilience of the agricultural systems under uncertain climate patterns. The environmental implications of these different policy priorities will probably be quite distinctive.

The long-term social acceptability of agri-environmental policy is likely to be influenced by the success of schemes in achieving their stated environmental goals, the cost of achieving these goals, and the significance of the reduction in food production that results from such schemes. There is already some criticism of the effectiveness of agri-environment schemes, although, because of the issues of scale discussed in Chapter 9, it will be difficult to definitively determine if schemes are working for some time. For related reasons the costs of achieving agri-environment scheme aims should decline over time, but current costs could be deemed unacceptable by the public if they were widely known. The final factor, security of food supply, is more likely to be influenced by climate change and political events than by the schemes themselves. Exactly what factors will influence the public's future willingness to pay for agri-environment schemes is unclear, but it is clear that in the very long term the public funding of farmers as 'wildlife gardeners' is unsustainable. In the past, environmental goods and services were the by-product of agricultural activity; with agri-environment schemes, they have become the product while food has become the by-product. The only truly sustainable future for agriculture and the environment is the development of viable production systems that are compatible with functioning ecosystems, in which the consumer of both pays the real cost.

International trade agreements (WTO)

International trade and trade liberalisation (the objective of trade agreements) has been shown to have an impact on the environment. Trade in agricultural commodities will play a greater role in meeting food needs of both developed and developing countries during the next decade, with competition among traditional exporters, primarily developed countries, intensifying and the emergence of new developing country exporters (OECD/FAO, 2005). Research has indicated that liberalisation of trade can facilitate environmental improvements and/or exacerbate environmental degradation. The global trade agreements, in particular the WTO, are having a significant impact on the development of agriculture policy by restricting programme options. It was recognised by WTO member countries that agri-environmental policies can alter production and price levels, which can influence trade patterns. For example, policies that provide financial support to encourage the adoption of environmentally benign production technology can have a large impact on production. The WTO negotiations will continue to pressure policy makers to complete the decoupling process, remove remaining trade barriers and reduce domestic subsidies affecting international trade (Dobbs and Pretty, 2001). In the WTO's Uruguay Round Agreement on Agriculture, it was recognised that countries can pursue domestic policy objectives, including the need to protect the environment and conserve natural resources. Agriculture policies must be decoupled from commodity production, including environmental payments, to be considered 'green box' and therefore not subject to trade action (see Chapter 2). Limits imposed by trade agreements may give greater prominence to 'green box' agri-environmental programmes as vehicles for farm income support although at present they play a very minor role. For example, environmental payments in the green box represented only 4.5% of the expenditures on agricultural support in OECD countries in 1998 (Diakosavvas, 2003). It has been noted that from the WTO's perspective a key policy concern is to distinguish between agri-environmental measures that actually address environmental issues, including limiting environmental costs and ensuring the provision of public goods associated with agriculture, from policies that appear to be merely labelled 'green' and used as a means of disguised protection (Diakosavvas, 2003). However, due to the 'multifunctional' nature of agri-environmental systems and their joint production of agricultural and environmental outputs, policies with environmental objectives may also influence production levels and trade flows. Therefore, the classification of policies as 'green box' may be very contentious. More recently, negotiations in the Doha Round (2004) established agreement of a framework for the elimination of export subsidies and the reduction of trade distorting domestic support by 20% in the first year, and substantial tariff reductions, while developing countries will benefit from special and differential treatment. However the specifics of this framework are still in negotiation. The future character of agri-environmental systems is likely to be significantly influenced by the degree of trade liberalisation and the role that agri-environmental policies can play in domestic agricultural support.

Multilateral environmental agreements

Multilateral (international) environmental issues and concerns appear to be emerging as an important influence on future agri-environmental policy. In the last 10 to 15 years a number of such environmental agreements have been negotiated and signed. In these agreements the signatory countries commit to meeting specific environmental objectives that will be influenced by agricultural development. For example, the 1997 Kyoto Protocol commits member countries to greenhouse-gas emission reduction targets for the period 2008–12. As discussed earlier, agriculture is both a source and a potential sink for greenhouse gases and as such will probably play an important role in many countries' emission reduction strategies. The Convention on Biological Diversity commits signatory countries to develop strategies for the conservation and sustainable use of biological diversity, including both natural or wild biodiversity and agricultural biodiversity – a subset of biodiversity essential to satisfy basic human needs for food security including heritage crop varieties and livestock breeds. For example, the European Biodiversity Action Plan develops measures, including agri-environmental measures, specifically aimed at supporting agricultural practices to conserve biodiversity (CEC, 2001). Other existing multilateral environmental agreements that may influence agri-environmental policy include:

- 1971 Ramsar Convention on Wetlands of International Importance
- 1979 Bonn Convention on the Conservation of Migratory Species
- 2001 Stockholm Convention on Persistent Organic Pollutants
- 2001 Cartagena Protocol on Biosafety

These existing and future multilateral environmental agreements will have an influence on agri-environmental policies and on the nature of agri-environmental systems in the future.

Summary

The nature of agriculture and the impact of agriculture on the environment is determined by forces and incentives that are often external to the farmer, who makes the fundamental management decisions. Factors such as input and output prices, as determined by market forces; policy initiatives at the local, regional, national and international level; technological trends and climate change all influence the management decisions of farmers. Further, all of these factors (and others) have a high degree of uncertainty and operate interdependently. As a result, it is impossible to accurately predict the nature of agri-environmental systems in even the near future. However, by recognising and understanding these contextual factors the predominant trends in these systems can be anticipated and, if required, addressed through various measures.

Throughout this book we have characterised agriculture as the applied science of diverting the earth's resources towards humans as food, fibres, fuel and other materials. Any such human-induced change can be considered as an environmental impact of agriculture. Over time (Chapter 3) we have learnt that many of these environmental impacts have had unforeseen, often negative, implications for humans. These include: direct poisoning of humans and wild-life by agro-chemicals, environmental pollution, a contribution to climate change, damage to ecological services such as flood defences and water purification, and declines in species and habitats valued for cultural and aesthetic reasons. Addressing these problems has been the driver for developing the range of policies and farming methods described in Chapters 2, 4, 5 and 8 that are designed to continue to supply our human needs but without jeopardising the wider environment and its capacity to provide for human needs into the future. In this final chapter we have seen that over time this balancing act of sustainable production is going to become both easier and more difficult with greater levels of uncertainty.

With advances in our understanding of both large-scale and fine-scale ecology we are starting to develop a better idea of how to balance human requirements for food, fibre and fuels with the long-term sustainability of the agro-ecosystem. Agricultural policies can be designed to support and encourage a multifunctional industry that produces foods, fibres and fuels from a healthy functioning ecosystem. Advances in technology will help us better identify and ameliorate the undesirable environmental impacts that result from future agricultural activity. In contrast, as the human population continues to increase in size and affluence, agriculture will be required to divert more of the world's resources into humans with greater pressure being placed on the environment. This situation can only be worsened by the uncertainty of climate change, which will require increased agricultural production in many areas to compensate for likely crop failures in others. This increased production is likely to result in greater selection pressure on other species, driving them to adapt to exploit this vast resource. Over time, this evolutionary struggle between agriculturalists and pest species will become more intense. But now we have a better understanding of how to identify and control pests while encouraging other species. Here again our ability to deal with the problem will be advanced by technology, but may be limited by market forces.

As a result of these conflicting factors the future nature of farming is uncertain; never again will it be considered purely as a system of generating food. It is clear that we are now beginning to recognise the significance of the multi-functionality of agricultural production. With increased environmental awareness the changes that occur in the future will be introduced with careful consideration of their likely wider environmental implications.

Glossary

AAPS	Arable Area Payments Scheme.
Agri-environment indicators	Measures used to monitor environmental changes within agricultural landscapes (e.g. soil quality, water quality, biodiversity).
Agri-environment scheme	Agricultural grant schemes, under which farmers receive payments for environmental enhancements.
Annual weed	Pest plant species with yearly (or less) life-cycles.
APF	Agriculture Policy Framework. Canada's federal agricultural policy approach that includes some agri-environmental policy.
Arable land	Land cultivated on an annual basis to grow crops for human consumption.
Artificial fertilisers	Chemical fertilisers derived from inorganic sources by industrial processing.
Beetle-bank	Grass strip planted in or at the edge of an arable field to encourage beetles and other invertebrate generalist predators.
Biodiesel	A processed fuel derived from biological sources such as vegetable oil (e.g. rapeseed, canola, soybean) that can readily be used in diesel engines.
Biodiversity	The variability among living organisms from all sources, including ecosystems and the ecological complexes of which they are part, including diversity within species, between species and of ecosystems.
BMP	Beneficial or best management practice. Management practices identified to minimise the impact of production systems or provide environmental benefits. BMPs are often supported as part of an agri-environmental scheme.
CAP	Common Agricultural Policy of the European Union.
Carbon sinks	A stock of carbon that is stored, often as organic carbon in soils and standing biomass (e.g. trees, shrubs, grass).

	Recognised in the Kyoto protocol as a mechanism to decrease atmospheric stocks of carbon as carbon dioxide, a greenhouse gas.
CBD	Convention on Biological Diversity, an international environmental agreement committing signatory countries to develop national plans for the preservation of biodiversity.
Cereals	Annual grass species cultivated for their seeds, e.g. wheat, oats, barley, etc.
Commodity	An undifferentiated product whose value arises from the owner's right to sell in an economic market (e.g. wheat, beef).
Common grazing	Area of usually unimproved hill ground over which several farmers have rights to graze.
Conservation headland	Outer strip of arable land cultivated in a way to encourage uncompetitive weeds, non-pest invertebrates and birds.
Conservation management	Agricultural management practices that conserve natural resources including soil, water, biodiversity.
Coppice and Coppicing	Method of harvesting trees for young stems by cutting to ground level on a rotational basis. Associated with elevated light levels and thus increased botanical diversity.
Coupled programme	Where there is a direct link between the level of agricultural production and the quantity of programme benefits paid to the farmer, and as such influences production.
Cross-compliance	A mechanism used in agri-environmental programmes whereby environmental conditions are attached to agricultural support policies such that the eligibility of a farmer or the level of support received is dependent on the meeting of specific environmental standards.
CRP	Conservation Reserve Program. A US set-aside programme that provides annual payments to farmers who voluntarily retire environmentally critical lands from crop production for ten years.
CSP	Conservation Security Program. A US agricultural programme that provides financial incentives to farmers to adopt or maintain practices that address soil, water and wildlife concerns.
Decoupled programme	Where programme payments and benefits do not depend on farmers' production choices, output levels or market conditions and do not influence production.
Deficiency payments	A combination of price and income supports paid to increase farmers' incomes by compensating for low commodity prices.

Defra	Department for Environment, Food and Rural Affairs (UK Government department).
Dry-stone wall and dry-stone dyke and dry-stain dyke	Field boundary constructed of stones without mortar. Several regional variants of structure and name. Considered important habitat for many species and of cultural/landscape importance.
Easement	The right of use of property (often land) for a specific purpose for an agent (individual, government, organisation) other than the landowner.
Eco-labels	A labelling system for consumer products, including food, that identifies the product as having some environmentally beneficial attribute. Often eco-labels identify some aspect of the production practice that would not be obvious to the consumer (e.g. organic food).
Ecological goods and services	Benefits provided to humans from ecosystems including water quality, water quantity, air purification, soil fertility, biodiversity benefits, etc.
Economic efficiency	An economically efficient allocation of resources is one that provides the greatest net benefits to society; or, alternatively, welfare cannot be increased by allocating goods and services in a different way.
Envirofund	Australian Government funding programme providing financial support to communities who undertake environmental conservation activities.
Environmental Farm Plan	A survey of all environmental features and environmental risks on a farm. Often used to direct or target environmental programmes.
EQIP	Environmental Quality Incentives Program. A US agricultural programme providing financial incentives to farmers for specific conservation management.
ESA	Environmentally Sensitive Area, early UK agri-environment scheme, restricted to target areas. Now defunct in Wales and Scotland.
Ethanol	Alcohol made from grain that can be used as a fuel for motor car either alone in a special engine or more commonly as an additive to petrol for petroleum engines.
Externality	Occurs when an individual or firm takes an action but does not bear all the costs (external cost or negative externality) or receive all the benefits (external benefit or positive externality).
FAO	Food and Agriculture Organization, an agency of the United Nations that works to improve levels of nutrition, standards of living and to eliminate hunger.

Field-margin	The edge of a field, usually considered as the grass strip at the edge of an arable field.
Flood meadow	Term frequently misused. A traditional form of encouraging spring grass growth for lambs, by flooding meadows in late winter.
Fodder	Any plant material grown to be feed to livestock.
GATT	General Agreement on Tariffs and Trade. A 1947 agreement designed to provide an international forum that encouraged free trade between member states by regulating and reducing tariffs on traded goods and providing mechanisms to resolve disputes. Succeeded by the WTO.
GIS	A Geographic Information System is a computer database to create, store, analyse and manage spatial data as well as displaying geographically referenced data.
GMO	A genetically modified organism is an organism whose genetic material has been altered using recombinant DNA technology. Often used to create crops with specific traits that could not have been developed using traditional crossbreeding.
'Good Farming Practice' (GFP)	Farmers in receipt of CAP agri-environmental scheme payments must comply with GFP codes that incorporate country-specific standards for water pollution, air pollution, fertiliser and pesticide use, etc.
GPS	Global Positioning System, a navigation system using a network of satellites.
Green box subsidies	As defined by the WTO, are subsidies that do not distort trade, are government funded and cannot involve price support. Are permitted under WTO rules and can involve environmental protection.
Grouse and grouse moor	Game bird of the uplands; moor managed by cyclical burning and grazing to encourage heather growth for grouse.
Habitat	Place where an organism lives.
Heather and heather moor/moorland	Dwarf shrubby species of the genus *Calluna* or *Erica*, may be most abundant species in upland unimproved land.
Hedge and hedgerow	Field boundary typically comprising a row of thorny short trees such as hawthorn and blackthorn.
Herbicide	Chemical applied to control pest plant species.
Hill farming	Upland agriculture typically associated with extensive grazing of sheep and beef, with the trend being for increased sheep and decreased cattle numbers.
Hill fence or hill dyke	The field boundary that separates hill ground from inbye land.

Hill ground or outbye	Large expanse of rough grazing, which is open ranched.
Improved grassland	Grassland whose composition has been altered to favour more productive and palatable species of grass and legumes. This may occur by reseeding, and/or applications of fertiliser and lime and/or grazing management.
Inbye	Lower enclosed ground (usually improved) surrounding the farm buildings. (Northern English/Scottish term) opposite to hill ground or outbye.
Insecticide	Chemical applied to control pest invertebrates.
IPCC	Intergovernmental Panel on Climate Change.
Jointness in production	Where a firm produces two or more outputs that are interlinked such that an increase or decrease in the supply of one affects the levels of the others.
Kyoto Protocol	An international agreement where ratifying countries commit to reduce their emissions of greenhouse gases by a target amount by the year 2012.
Landcare	Australian agricultural programme that provides financial and institutional support for community-based initiatives to address environmental issues.
Land drain and drainage	Pipe buried in agricultural land to improve the drainage.
Ley	Typically relatively short-term improved grass or grass/clover pasture.
Lowland heath	Heather-dominated vegetation in the lowlands associated with nutrient-poor soils. May be dry in the east, wet in the west or coastal regions.
Manure and FYM	Farm yard manure, solid waste products of stock, including faeces, urine and bedding, used as organic fertiliser.
Market failure	Occurs when an economic market does not lead to an allocation of resources that is best for society.
Meadow	Grass or grass/herb vegetation, usually cut for hay or silage in summer and grazed at other times.
Monoculture	The cultivation of a single species of crop within a field. Does not mean growing the same crop year after year, although this may occur.
Mowing and topping	Cutting grass field typically for winter feed or to remove rank stems or unwanted species.
Multifunctional agriculture	Agricultural industry that produces both food and fibre commodities as well as a range of non-commodity outputs including environmental benefits, aesthetic and social values.
Natural regeneration	Process of vegetation succession typically involving the invasion of scrub and trees.

NEGTAP	National Expert Group on Transboundary Air Pollution.
Non-market goods and services	Goods and services that are not traded in an economic market and as a result do not have prices reflecting value to society. Often include environmental goods and services such as biodiversity, hydrological services, etc.
Non-point-source pollution	Pollution that comes from many dispersed sources that are difficult or impossible to identify making it difficult to regulate.
OECD	Organisation for Economic Co-operation and Development.
Pasture	Grass or grass/herb vegetation, usually seasonally grazed.
Pesticide	Any substance or mixture of substances intended for preventing, destroying, repelling, or mitigating any pest including insect, weed, fungal, bacterial, bird or mammal pests.
Pollard	Method of harvesting trees for young stems by cutting to head height on a rotational basis. Thus preventing stock from browsing the regrowth.
Price	In economics and business, price presents the value of a good or service to an individual or society and is determined by the interaction of consumers and producers in an economic market.
Public good	A good that is non-rival (one individual's consumption does not change the quantity or quality available for other consumers) and non-excludable (is expensive or impossible to prevent an extra individual from enjoying). Due to these characteristics public goods are not traded in a market, are not provided by profit-maximising firms and are often a cause of market failures.
Rough grazing	Unimproved grassy vegetation, dominated by non-agricultural species, frequently found in the uplands.
SAC	Special Area of Conservation. European designation of legally protected land because of its conservation value.
Scrub	Area of land covered with low-growing trees and shrubs, such as gorse, hawthorn or juniper.
Semi-improved grassland	Grassland that has received some application of fertiliser or lime, but whose composition is more diverse than just rye-grass and clover. May also be derived from a reverting improved pasture.
Set-aside	Land taken out of arable production, receiving government funding to reduce production.
Silage	Method of storing fodder for winter feed, without drying.
Slurry	Semi-solid animal waste products. Produced during housing of stock.

Sodbuster	A provision of US agriculture policy that requires farmers who convert highly erodible cropland to annual cultivation to do so under an approved erosion control plan or forfeit eligibility for other government farm programme benefits.
Spray drift	Unintentional drift of agrochemical into non-target area.
Stewardship scheme and Countryside Stewardship scheme	Agri-environment schemes in England and Scotland.
Stubble	Residues of stems and spilt grain following harvesting of an arable crop.
Supplementary feeding	The feeding of extra fodder to stock grazing fresh pasture.
Swampbuster	A provision of US agriculture policy that requires all farmers to protect wetlands on their land in order to be eligible for other government farm programme benefits.
Urbanisation	Process whereby an increasing proportion of the total population of a region lives in urban areas (cities and towns).
USDA	United States Department of Agriculture.
Weeds and weedkillers	Undesirable plant species and the method of their chemical control.
Wetlands	Area of impeded drainage, often of conservation value. May also be important as flood defence.
Wild flowers	Within an agricultural context, these typically include perennial species of unimproved grasslands and moorlands plus some annual species of arable fields.
Wildlife corridor	Linear feature crossing agricultural land such as stream or hedge which is said to encourage the movement of wildlife across open fields.
WHIP	Wildlife Habitat Incentives Program. A US agriculture programme providing financial assistance to farmers to reclaim and conserve wildlife habitat on productive land.
WQIP	Water Quality Incentive Program. A US agriculture programme providing financial assistance to farmers to conserve water quality.
WRP	Wetland Reserve Program. A US agriculture programme providing financial assistance to farmers to create and conserve wetlands.
WTO	World Trade Organization. Global organisation dealing with the rules of trade between nations with an objective to ensure that trade flows smoothly, predictably and freely as possible. The successor to the GATT.

References

Adams, R. M., Hurd, B. H. and Reilly, J. (1999). *A Review of Impacts to U.S. Agricultural Resources*. Prepared for the Pew Center on Global Climate Change. February, 1999. www.pewclimate.org/docUploads/env%5Fargiculture%2Epdf.

Aerts, R., Berendse, F., DeCaluwe, H. and Schmitz, M. (1990). Competition in heathland along an experimental gradient of nutrient availability. *Oikos*, **57**, 310–18.

Alonso, I. and Hartley, S. E. (1998). Effects of nutrient supply, light availability and herbivory on the growth of heather and three competing grasses. *Plant Ecology*, **137**, 203–12.

Andrewartha, H. G. and Birch, L. C. (1954). *The Distribution and Abundance of Animals*. Chicago: University of Chicago Press.

Aspinall, R. and Pearson, D. (2000). Integrated geographical assessment of environmental condition in water catchments: linking landscape ecology, environmental modelling and GIS. *Journal of Environmental Management*, **59**, 299–319 Sp. Iss. SI AUG 2000.

Australian Government, Department of Agriculture, Fisheries and Forestry Website, Landcare Program. www.affa.gov.au/content/output.cfm?ObjectID=D2C48F86-BA1A-11A1-A2200060B0A04284&contType=outputs, Accessed January, 2006.

Bakker, J. P. (1989). *Nature Management by Grazing and Cutting*. Dordrecht: Kluwer Academic Publishers.

Baldock, D. (1990). *Agriculture and Habitat Loss in Europe*. WWF International CAP Discussion Paper No. 3. Published by WWF.

Bardgett, R. D., Marsden, J. H. and Howard, D. C. (1995). The extent and condition of heather on moorland in the uplands of England and Wales. *Biological Conservation*, **71**, 155–61.

Baylis, K., Raiser, G. and Simon, L. (2003). Agri-Environment Programs and the Future of the WTO. *Proceedings of the International Conference – Agriculture Policy Reform and the WTO: Where are we Heading*? Capri, Italy, June 23-6, 2003. Camberley, Surrey: Edward Elgar Publishing Ltd.

Bender, D. J., Tischendorf, L. and Fahrig, L. (2003). Using patch isolation metrics to predict animal movement in binary landscapes. *Landscape Ecology*, **18**, 17-39.

Bengtsson, J., Ahnstom, J. and Weibull A. C. (2005). The effects of organic agriculture on biodiversity and abundance: a meta-analysis. *Journal of Applied Ecology*, **42**, 261-9.

Berendse, F., Oomes, M. J. M., Altena, H. J. and Elberse, W. Th. (1992). Experiments on the restoration of species-rich meadows in the Netherlands. *Biological Conservation*, **62**, 59-65.

Bezemer, T. M., Lawson, C. S., Hedlund, K. *et al.* (2006). Plant species and functional group effects on abiotic and microbial soil properties and plant–soil feedback responses in two grasslands. *Journal of Ecology*, **94**, 893-904.

Bhattacharya, M., Primack, R. B. and Gerwein, J. (2003). Are roads and railroads barriers to bumblebee movement in a temperate suburban conservation area? *Biological Conservation*, **109**, 37-45.

Bignal, E. and Baldock, D. (2002). Agri-environmental policy in a changing European context. In *Conservation Pays? Reconciling Environmental Benefits with Profitable Grassland Systems*, ed. J. Frame. Occasional Symposium No. 36 British Grassland Society, pp. 3-14.

Bignal, E. M. and McCracken, D. I. (1996). Low-intensity farming systems in the conservation of the countryside. *Journal of Applied Ecology*, **33**, 413-24.

Biodynamic Farming Association. (2006). www.biodynamics.com/biodynamics.html.

Bischoff, A., Crémieux, L., Smilauerova, M. *et al.* (2006). Detecting local adaptation in widespread grassland species – the importance of scale and local plant community. *Journal of Ecology*, **94**, 1130-42.

Blackstock, T. H., Rimes, C. A., Stevens, D. P. *et al.* (1999). The extent of semi-natural grassland communities in lowland England and Wales: a review of conservation surveys 1978-96. *Grass and Forage Science*, **54**, 1-18.

Bobbink, R., Hornung, M. and Roefofs, J. G. M. (1998). The effects of air-borne nitrogen pollutants on species diversity in natural and semi-natural European vegetation. *Journal of Ecology*, **86**, 717-38.

Boetsch, J. R., Van Manen, F. K. and Clark, J. D. (2003). Predicting rare plant occurrence in Great Smoky Mountains National Park. *USA Natural Areas Journal*, **23**, 229-37.

Bowers, J. K. (1985). British agricultural policy since the Second World War. *The Agricultural History Review*, **33**, 66-76.

Bracken, F. and Bolger, T. (2006). Effects of set-aside management on birds breeding in lowland Ireland. *Agriculture Ecosystems and Environment*, **117**, 178-84.

Buckwell, A. (1997). *Towards a Common Agricultural and Rural Policy for Europe*. European Economy – Reports and Studies No. 5, Brussels, Belgium. http://ec.europa.eu/agriculture/publi/buck_en/cover.htm.

Bunyan, J. P. and Stanley, P. I. (1983). The environmental cost of pesticide usage in the United Kingdom. *Agriculture, Ecosystems and Environment*, **9**, 197-209.

Cain, Z. and Lovejoy, S. (2004). History and outlook for farm bill conservation programs. *Choices Magazine*. www.choicesmagazine.org.

Carpentier, C. L., Bosch, D. J. and Batie. S. S. (1998). Using spatial information to reduce costs of controlling agricultural nonpoint source pollution. *Agricultural Resource Economics Review*, **27**, 72–84.

Carson, R. (1962). *Silent Spring*. London: Penguin (1991 Printing).

Chamberlain, D. E., Fuller, R. J., Bunce, R. G. H., Duckworth, J. C. and Shrubb, M. (2000). Changes in the abundance of farmland birds in relation to the timing of agricultural intensification in England and Wales. *Journal of Applied Ecology*, **37**, 771–88.

Christie, M., Hanley, N., Warren, J. *et al.* (2006). Valuing the diversity of biodiversity. *Ecological Economics*, **58**, 304–17.

Claasen, R., Hansen, L., Peters, M. *et al.* (2001). *Agri-Environmental Policy at the Crossroads: Guideposts on a Changing Landscape*. United States Department of Agriculture, Agricultural Economic Report Number 794. Washington, DC.

Commission of the European Communities (CEC). (2001). *Communication from the Commission to the Council and the European Parliament: Biodiversity Action Plan for Agriculture*. COM (2001)162 final, Volume III. Brussels, 27.3.2001.

Commission of the European Communities (CEC). (2002). *Communication from the Commission to the Council and the European Parliament: Implementation of the Council Directive 91/676/EEC Concerning the Protection of Water Against Pollution Caused by Nitrates from Agricultural Sources*. COM (2002) 407.

Common, M. and Perrings, C. (1992). Towards an ecological economics of sustainability. *Ecological Economics*, **6**, 7–34.

Crawford, T. J. and Jones, D. A. (1988). Variation in the colour of the keel petals in *Lotus corniculatus* L. 4. Morph distribution in the British Isles. *Heredity*, **61**, 175–88.

Davenport, I. J., Wilkinson, M. J., Mason, D. C. *et al.* (2000) Quantifying gene movement from oilseed rape to its wild relatives using remote sensing. *International Journal of Remote Sensing*, **21**, 3567–73.

De Deyn, G. B., Raaijmakers, C. E. and van der Putten, W. H. (2004). Plant community development is affected by nutrients and soil biota. *Journal of Ecology*, **92**, 824–34.

Department for Environment, Food and Rural Affairs (Defra). (2002a). *Agricultural and Horticultural Census for England 1980*. www.statistics.defra.gov.uk/esg/default.asp/.

Department for Environment, Food and Rural Affairs (Defra). (2002b). *Agricultural and Horticultural Census for England 1990*. www.statistics.defra.gov.uk/esg/default.asp/.

Department for Environment, Food and Rural Affairs (Defra). (2002c). *Ammonia in the United Kingdom*. www.defra.gov.uk/corporate/publications/pubcat/index.htm.

Department for Environment, Food and Rural Affairs (Defra). (2002d). *The British Survey of Fertiliser Practice*. www.defra.gov.uk/farm/environment/land-manage/nutrient/pdf/bsfp2002.pdf.

Department for Environment, Food and Rural Affairs (Defra). (2002e). *Using Economic Instruments to Address the Environmental Impacts on Agriculture*. www.defra.gov.uk/farm/sustain/newstrategy/econ/section2.pdf.

Department for Environment, Food and Rural Affairs (Defra). (2003). *Agricultural and Horticultural Census for England 1920*. www.statistics.defra.gov.uk/esg/default.asp/.

Department for Environment, Food and Rural Affairs (Defra). (2004). *Strategic Review of Diffuse Water Pollution from Agriculture. Initial Appraisal of Policy Instruments to Control*

Water Pollution from Agriculture. www.defra.gov.uk/farm/environment/water/csf/previous-papers.htm.

Department for Environment, Food and Rural Affairs (Defra). (2005). *Managing GM Crops with Herbicides: Effects on Farmland Wildlife.* www.defra.gov.uk/environment/gm/fse/results/fse-summary-05.pdf.

Department for Environment, Food and Rural Affairs (Defra). (2006). *Bio-energy Infrastructure Scheme: Explanatory Booklet.* www.defra.gov.uk/farm/acu/energy/infrastructure-booklet.pdf.

Diakosavvas, D. (2003). The greening of the WTO Green Box: a quantitative appraisal of agri-environmental policies in OECD countries. *Proceedings of the International Conference–Agricultural Policy Reform and the WTO: Where are we Heading?* Capri, Italy, June 23–6, 2003. Camberley, Surrey: Edward Elgar Publishing Ltd.

Dobbs, T. and Pretty, J. (2001). *Future Directions for Joint Agricultural-Environmental Policies: Implications of the United Kingdom Experience for Europe and the United States.* South Dakota State University Economics Research Report 2001-1 and University of Essex Centre for Environment and Society Occasional Paper 2001-5. August 2001.

Dodd, M., Silverton, J., McConway, K., Potts, J. and Crawley, M. (1995). Community stability – a 60 year record of trends and outbreaks in the occurrence of species in the Park Grass Experiment. *Journal of Ecology*, **83**, 277–85.

Eco-Labels Website (2006). www.eco-labels.org/home.cfm, accessed March 2006.

Edwards, A. R., Mortimer, S. R., Lawson, C. S. *et al.* (2007) Hay strewing, brush harvesting of seed and soil disturbance as tools for the enhancement of botanical diversity in grasslands. *Biological Conservation*, **134**, 372–82.

Effland, A. B. W. (2000). US farm policy: the first 200 years. *Agricultural Outlook*, USDA, **269**, 21–5. See also www.ers.usda.gov/publications/agoutlook/mar2000/ao269g.pdf.

Egler, F. E. (1954). Vegetation science concepts: I. Initial floristic composition, a factor in old-field vegetation development. *Vegetatio*, **4**, 412–17.

European Commission. (1997). *Report to the European Parliament and to the Council on the Application of Council Regulation (EEC) No. 2078/92 on Agricultural Production Methods compatible with the requirements of the protection of the environment and the maintenance of the countryside.* European Commission Directorate-General for Agriculture.

European Commission. (2002a). *Agriculture in the European Union – Statistical and Economic Information 2001.* European Commission Directorate-General for Agriculture.

European Commission. (2002b). *EAAGF Financial Report.* European Commission Directorate-General for Agriculture.

European Commission. (2003a). *Agriculture in the European Union – Statistical and Economic Information 2002.* European Commission Directorate-General for Agriculture.

European Commission. (2003b). *Rural Development Programmes, 2000–2006.* European Commission Directorate-General for Agriculture.

European Commission. (2003c). *European Union and the Common Agricultural Policy. Public Opinion in the Member States. Special Eurobarometer 190.* European Commission Directorate-General for Agriculture.

European Commission. (2003d). *Directorate-General for Agriculture, Agriculture and the Environment: Fact Sheet*. http://europa.eu.int/comm/agriculture/publi/fact/envir/2003_en.pdf.

European Commission. (2006). *Agriculture and Environment*. http://europa.eu.int/comm/agriculture/envir/index_en.htm, accessed March 2006.

European Court of Auditors. (2000). *Special Report no 14/00 on Greening the CAP*.

Farming Statistics Team, Defra. (2002–2004). *Agricultural and Horticultural Census for England 1900–2000*. www.statistics.defra.gov.uk/esg/default.asp/.

Feather, P., Hellerstein, D. and Hansen, L. (1999). *Economic Valuation of Environmental Benefits and the Targeting of Conservation Programs: The Case of the CRP*. AER-778, US Department of Agriculture Economic Research Service, April.

Fischer, G., Shah, M., Tubiello, F. N. and van Velhuizen, H. (2005). Socio-economic and climate change impacts on agriculture: an integrated assessment, 1990–2080. *Philosophical Transactions of The Royal Society B*, **360**, 2067–83.

Flora, C. B. (1995). Sustainability: agriculture and communities in the great plains and corn belt. *Research in Rural Sociology and Development*, **6**, 227–46.

Ford, E. B. (1964). *Ecological Genetics*. London: Chapman and Hall.

Fresco, L. O. (2002). *The Future of Agriculture: Challenges for Environment, Health and Safety Regulation of Pesticides*. From a presentation made to the OECD Working Group on Pesticides, Paris. February 4, 2002.

Fuller, R. J., Gregory, R. D., Gibbons, D. W. *et al.* (1995). Population declines and range contractions among farmland birds in Britain. *Conservation Biology*, **9**, 1425–41.

Gimingham, C. H. (1972). *Ecology of Heathlands*. London: Chapman and Hall, New York: Banes & Noble.

Gough, M. W. and Marrs, R. H. (1990). A comparison of soil fertility between semi-natural and agricultural plant communities: implications for the creation of species-rich grassland on abandoned agricultural land. *Biological Conservation*, **51**, 83–96.

GreatorexDavies, J. N., Hall, M. L. and Marrs, R. H. (1992). The conservation of the pearl-boarded fritillary butterfly (*Boloria euphrosyne* L.). Preliminary studies on the creation and management of glades in conifer plantations. *Forest Ecology and Management*, **53**, 1–14.

Green, B. H. (1990). Agricultural intensification and the loss of habitat, species and amenity in British grasslands: a review of the historical change and assessment of future prospects. *Grass and Forage Science*, **45**, 365–72.

Haddad, N. M. (1999). Corridor and distance effects on interpatch movements: a landscape experiment with butterflies. *Ecological Applications*, **9**, 612–22.

Halberg, N., Verschuur, G. and Goodlass, G. (2005). Farm level environmental indicators; are they useful? An overview of green accounting systems for European farms. *Agriculture Ecosystems and Environment*, **105**, 195–212.

Hald, A. B. (1999). The impact of changing season in which cereals are sown on the diversity of the weed flora in rotational fields in Denmark. *Journal of Applied Ecology*, **36**, 24–32.

Hamilton, G. (2000). When good bugs turn bad. *New Scientist*, **15**, January 2000, 31–3.

Hancock, J. F. (2005). Contributions of domesticated plant studies to our understanding of plant evolution. *Annals of Botany*, **96**, 953–63.

Hansen, L. (2006). Wetlands status and trends. In *Agricultural Resources and Environmental Indicators*, 2006 edn, ed. K. Wiebe and N. Gollehon. United States Department of Agriculture. Economic Information Bulletin No. (EIB-16), July 2006.

Hargrove, W. W., Gardner, R. H., Turner, M. G., Romme, W. H. and Despain, D. G. (2000). Simulating fire patterns in heterogeneous landscapes. *Ecological Modelling*, **135**, 243–63.

Haynes, R. and Williams, P. (1993). Nutrient cycling and soil fertility in the grazed pasture ecosystem. *Advances in Agronomy*, **49**, 119–99.

Heil, G. W. and Diemont, W. H. (1983). Raised nutrient levels change heathland into grassland. *Vegetatio*, **53**, 113–20.

Hole, D. G., Perkins, A. J., Wilson, J. D. *et al.* (2005). Does organic farming benefit biodiversity? *Biological Conservation*, **122**, 113–30.

Hopkins, J. J. (1989). Prospects for habitat creation. *Landscape Design*, **179**, 19–23.

Huggard, D. J. (2003). Use of habitat features, edges and harvest treatments by spruce grouse in subalpine forest. *Forest Ecology and Management*, **175**, 531–44.

Ikerd, J. (2006). www.sustainable-ag.ncsu.edu/whatissa/definesa.htm.

Intergovernmental Panel on Climate Change (IPCC). (1997). *Summary for Policymakers: The Regional Impacts of Climate Change: An Assessment of Vulnerability*, ed. R. T. Watson, M. C. Zinyowera, R. H. Moss and D. J. Dokken. A Special Report of IPCC Working Group II. November 1997.

Intergovernmental Panel on Climate Change (IPCC). (2000). *Summary for Policymakers: Land Use, Land-Use Change, and Forestry*. A Special Report of the Intergovernmental Panel on Climate Change. Approved in detail at IPCC Plenary XVI, Montreal, Canada, May 2000. www.ipcc.ch/pub/SPM_SRLULUCF.pdf.

Intergovernmental Panel on Climate Change (IPCC). (2001). *Climate Change 2001. Third Assessment Report of the Intergovernmental Panel on Climate Change*. Cambridge: Cambridge University Press.

Jepsen, J. U., Topping, C. J., Odderskaer, P. and Andersen, P. N. (2005). Evaluating consequences of land-use strategies on wildlife populations using multiple-species predictive scenarios. *Agriculture Ecosystems and Environment*, **105**, 581–94.

Kay, D., Wyer, M. D., Crowther, J. *et al.* (2005). Sustainable reduction in the flux of microbial compliance parameters from urban and arable land use to coastal bathing waters by a wetland ecosystem produced by a marine flood defence structure. *Water Research*, **39**, 3320–32.

Lampkin, N. L. (1997). Organic livestock production and agricultural sustainability. In *Resource Use in Organic Farming, Proceedings of the Third ENOF Workshop*; Ancona, pp. 321–30.

Latacz-Lohmann, U. and Hodge, I. (2003). European agri-environmental policy for the 21st century. *The Australian Journal of Agricultural and Resource Economics*, **47**, 123–9.

Lawson, C. S., Ford, M. A. and Mitchley, J. (2004a). The influence of seed addition and cutting regime on the success of grassland restoration on ex-arable land. *Applied Vegetation Science*, **7**, 259–66.

Lawson, C. S., Ford, M. A., Mitchley, J. and Warren, J. M. (2004b). The establishment of heathland vegetation on ex-arable land: the response of *Calluna vulgaris* to soil acidification. *Biological Conservation*, **116**, 409–16.

Lehtonen, H., Lankoski, J. and Niemi, J. (2005). *Evaluating the Impact of Alternative Agricultural Policy Scenarios on Multifunctionality: A Case Study of Finland*. European Network of Agricultural and Rural Policy Research Institutes. Working Paper No.13, July 2005. www.enarpri.org.

Leopold Centre. (2006). http://www.leopold.iastate.edu/about/sustainableag.htm.

MacArthur, R. H. and Wilson, E. O. (1967). *The Theory of Island Biogeography*. Princeton, NJ: Princeton University Press.

Marrs, R. H. (1985). Techniques for reducing soil fertility for nature conservation purposes: a review in relation to research at Roper's Heath, Suffolk, England. *Biological Conservation*, **34**, 307–32.

Marrs, R. H. (1993). Soil fertility and nature conservation in Europe: theoretical considerations and practical management solutions. *Advances in Ecological Research*, **24**, 241–300.

Marrs, R. H., Williams, C. T., Frost, A. J. and Plant, R. A. (1989). Assessment of the effects of herbicide spray drift on a range of plant species of conservation interest. *Environmental Pollution*, **59**, 71–86.

Marrs, R. H., Snow, C. S. R., Owen, K. M. and Evans, C. E. (1998). Heathland and acid grassland creation on arable soils at Minsmere: identification of the potential problems and a test of cropping to impoverish soils. *Biological Conservation*, **85**, 69–82.

Mauremooto, J. R., Wratten, S. D., Worner, S. P. and Fry, G. L. A. (1995). Permeability of hedgerows to predatory carabid beetles. *Agriculture Ecosystems and Environment*, **52**, 141–8.

McCraken, D. I. and Tallowin, J. R. (2004). Swards and structure: the interactions between farming practices and bird food resources in lowland grasslands. *Ibis*, **146** (Suppl 2), 108–14.

McKenzie, D. F. and Riley, T. Z. (1995). *How Much is Enough? A Regional Wildlife Habitat Needs Assessment for the 1995 Farm Bill*. Wildlife Management Institute and Soil and Water Conservation Society. Feb.

McRae, T., Smith, C. A. S. and Gregorich, L. J. (eds.). (2000). *Environmental Sustainability of Canadian Agriculture: Report of the Agri-Environmental Indicator Project*. Ottawa, Ontario: Agriculture and Agri-Food Canada.

Mollinson, B. and Holmgren, D. (1978). *Permaculture One*. Tyalgum: Tagari

National Expert Group on Transboundary Air Pollution (NEGTAP). (2001). *Transboundary Air Pollution: Acidification, Eutrophication and Ground-Level Ozone in the UK*. Prepared on behalf of the UK Department for Environment, Food and Rural Affairs. London: The National Expert Group on Transbounday Air Pollution.

Nature Conservancy Council. (1986). *Potentially Damaging Operations Manual*. Peterborough: NCC.

Newman, E. I. (1997). Phosphorus balance of contrasting farming systems, past and present. Can food production be sustainable? *Journal of Applied Ecology*, **34**, 1334–47.

Ng, S. J., Dole, J. W., Sauvajot, R. M., Riley, S. P. D. and Valone, T. J. (2004). Use of highway undercrossings by wildlife in southern California. *Biological Conservation*, **115**, 499–507.

Northbourne, Lord (1940). *Look to the Land*. London: Dent.

Norton, B. G. (1987). *Why Preserve Natural Variety*. Princeton, NJ: Princeton University Press.

Noss, R. F. and Cooperrider, A. Y. (1994). *Saving Nature's Legacy: Protecting and Restoring Biodiversity*. Washington, DC: Island Press.

Organisation for Economic Co-operation and Development (OECD). (2001a). *Multifunctionality: Towards an Analytical Framework*. Paris: OECD.

Organisation for Economic Co-operation and Development (OECD). (2001b). *Environmental Indicators for Agriculture Volume 3: Methods and Results*. Paris: OECD.

Organisation for Economic Co-operation and Development (OECD). (2003a). *Agriculture and the Environment: Lessons Learned from a Decade of OECD Work*. Paris: OECD. www.oecd.org.

Organisation for Economic Co-operation and Development (OECD). (2003b). *Agri-Environmental Policy Measures: Overview of Developments*. Joint Working Party on Agriculture and the Environment. *COM/AGR/CA/ENV/EPOC(2002)95/FINAL. JT00153105.* November 6, 2003.

Organisation for Economic Co-operation and Development (OECD). www.oecd.org.

Organisation for Economic Co-operation and Development (OECD) and Food and Agriculture Organization (FAO) of the United Nations. (2005). *OECD-FAO Agricultural Outlook: 2005–201: Highlights, 2005*. www.oecd.org/dataoecd/32/51/ 35018726.pdf.

Owen, K. M. and Marrs, R. H. (2000). Creation of heathland on former arable land at Minsmere, Suffolk, UK – the effects of soil acidification on the establishment of *Calluna* and ruderal species. *Biological Conservation*, **93**, 9–18.

Payraudeau, S. and Van der Werf, H. M. G. (2005). Environmental impact assessment for a farming region: a review of methods. *Agriculture Ecosystems and Environment*, **107**, 1–19.

Permaculture International. (2006). http://www.permacultureinternational.org/ whatispermaculture.htm.

Permaculture Net. (2006). http://www.permaculture.net/.

Pinstrup-Andersen, P. (2001). *The Future World Food Situation and the Role of Plant Diseases*. The Plant Health Instructor. DOI:10.1094/PHI-I-2001-0425-01.

Pionke, H. B., Gburek, W., Sharpley, A. N. and Zollweg, J. A. (1997). Hydrologic and chemical controls on phosphorus loss from catchments. In *Phosphorus Loss to Water from Agriculture*, ed. H. Tunney. Cambridge, England: CAB International Press, pp. 225–42.

Pollard, E., Hopper, M. D. and Moore, N. W. (1974). *Hedges. New Naturalist Series 58*. London: Collins, 256pp.

Popper, D. E. and Popper, F. J. (1987). The great plains: from dust to dust. *Planning*. www.planning.org/25anniversary/planning/1987dec.htm.

Potash and Phosphate Institute (PPI). (2002). *Plant nutrient use in North America*. Technical Bulletin. 2002–1 Norcross, GA.

Potter, C. (1998). *Against the Grain: Agri-Environmental Reform in the United States and Europe*. Wallingford: CAB International Press.

Power, J. F. and Schepers, J. S. (1989). Nitrate contamination of groundwater in North America. *Agriculture, Ecosystems and Environment*, **26**, 165–87.

Pretty, J. N., Brett, C., Gee, D. *et al.* (2000). An assessment of the total external costs of UK Agriculture. *Agricultural Systems*, **65**, 113–36.

Pywell, R. F., Webb, N. R. and Putwain, P. D. (1994). Soil fertility and its implications for the restoration of heathland on farmland in Southern Britain. *Biological Conservation*, **70**, 169–81.

Pywell, R. F., Webb, N. R. and Putwain, P. D. (1997). The decline in heathland seed populations following the conversion to agriculture. *Journal of Applied Ecology*, **34**, 949–60.

Pywell, R. F., Bullock, J. M., Roy, D. B. *et al.* (2003). Plant traits as predictors of performance in ecological restoration. *Journal of Applied Ecology*, **40**, 65–77.

Radcliffe, D. A. (1984). Post-medieval and recent changes in British vegetation: the culmination of human influence. *New Phytologist*, **98**, 73–100.

Randall, A. (2002). Valuing the outputs of multifunctional agriculture. *European Review of Agricultural Economics*, **29**, 289–307.

Ratcliffe, D. (1984). *Nature Conservation Review of Great Britain*. Peterborough: NCC.

Read, H. J. (ed.) (1996). *Pollard and Veteran Tree Management II*. Corporation of London.

Ribaudo, M. (1997). Water quality programs. *Agricultural Resources and Environmental Indicators 1996–97*. AH-705, United States Department of Agriculture. Economic Research Services, July.

Rich, T. C. G. and Woodruff, E. R. (1996). Changes in vascular plant floras of England and Scotland between 1930–60 and 1987–88: the BSBI monitoring scheme. *Biological Conservation*, **75**, 217–29.

Robinson, R. A. and Sutherland, W. J. (2002). Post-war changes in arable farming and biodiversity in Great Britain. *Journal of Applied Ecology*, **39**, 157–76.

Robson, N. (1997). The evolution of the Common Agricultural Policy and the incorporation of environmental considerations. In *Farming and Birds in Europe: The Common Agricultural Policy and its Implications for Bird Conservation*, ed. D. J. Pain and M. W. Pienkowski. London: Academic Press, pp. 43–78.

Rodwell, J. S. (ed.) (1991a). *Woodlands and scrub. Vol. 1 of British Plant Communities*. Cambridge: Cambridge University Press.

Rodwell, J. S. (ed.) (1991b). *Mires and Heaths. Vol. 2 of British Plant Communities*. Cambridge: Cambridge University Press.

Rodwell, J. S. (ed.) (1992). *Grasslands and Montane Communities. Vol. 3 of British Plant Communities*. Cambridge: Cambridge University Press.

Rodwell, J. and Patterson, G. (1994). *Creating New Native Woodlands. Forestry Commission Bulletin 112*. London: HMSO.

Sanzenbacher, P. M. and Haig, S. M. (2002). Regional fidelity and movement patterns of wintering Killdeer in an agricultural landscape. *Waterbirds*, **25**, 16–25.

Schrøder, H. (1985). Nitrogen losses from Danish agriculture – trends and consequences. *Agriculture, Ecosystems and Environment*, **14**, 279–89.

Scottish Environment Protection Agency (SEPA). (2000). www.sepa.org.uk/pdf/ publications/leaflets/sapg/reviews/2000.pdf.

Sharpley, A. N., Daniel, T., Sims, T. J. *et al.* (2003). *Agricultural Phosphorus and Eutrophication*, 2nd edn. United States Department of Agriculture. Agricultural Research Service ARS-149.

Sheail, J. (1995). Nature protection, ecologists and the farming context: a UK historical perspective. *Journal of Rural Studies*, **11**, 79–88.

Silvey, V. (1986). The contribution of new varieties to cereal yields in England and Wales between 1947 and 1983. *Journal of the National Institute of Agricultural Botany*, **17**, 155–68.

Smil, V. (1997). Global population and the nitrogen cycle. *Scientific American*, **277**, 76–81.

Smith, R. S. and Rushton, S. P. (1994). The effects of grazing management on the vegetation of mesotrophic (meadow) grassland in northern England. *Journal of Applied Ecology*, **31**, 13–24.

Smith, R. S., Shiel, R. S., Millward, D. and Corkhill, P. (2000). The interactive effects of management on the productivity and plant community structure of an upland meadow: an 8-year field trial. *Journal of Applied Ecology*, **37**, 1029–43.

Smith, R. S., Shiel, R., Bardgett, R. D. *et al.* (2003). Soil microbial community, fertility, vegetation and diversity as targets in the restoration management of a meadow grassland. *Journal of Applied Ecology*, **40**, 51–64.

Soil Association. (2006). www.soilassociation.org.

Statistics Canada. (2005). www.statcan.ca/start.html.

Steiner, C. F., Long, Z. T., Krumins, J. A. and Morin, P. J. (2006). Population and community resilience in multitrophic communities. *Ecology*, **87**, 996–1007.

Stoate, C. (1995). The changing face of lowland farming and wildlife. Part 1 1845–1945. *British Wildlife*, **6**, 341–50.

Syphard, A. D., Clarke, K. C. and Franklin, J. (2005). Using a cellular automaton model to forecast the effects of urban growth on habitat pattern in southern California. *Ecological Complexity*, **2**, 185–203.

Thirsk, J. (1997). *Alternative Agriculture: a History From the Black Death to the Present Day*. Oxford: Oxford University Press.

Thomassin, P. J. (2003). Canadian agriculture and the development of carbon trading and offset system. *American Journal of Agricultural Economics*, **85**(5), 1171–7.

Tilman, D., Reich, P. B., Knops, J. *et al.* (2001). Diversity and productivity in a long-term grassland experiment. *Science*, **294**, 843–5.

Topping, C. J. (1999). An individual-based model for dispersive spiders in agroecosystems: simulations of the effects of landscape structure. *Journal of Arachnology*. **27**, 378–86.

United Nations Economic Commission for Europe (UNECE). (1999) Protocol to the 1979 Convention on Long Range Transboundary Air Pollution (CLRTAP) to Abate Acidification, Eutrophication and Ground Level Ozone. 1 December 1999. Gothenburg, Sweden.

United Nations Environment Programme (UNEP). (1997). *Global Environment Outlook-1: Global State of the Environment Report 1997*. www.grida.no/geo1/ch/ch4_14.htm.

United Nations Framework Convention on Climate Change (UNFCCC). (2006). *National Inventory Submissions 2006 on Greenhouse Gases*. http://unfccc.int/.

United States Department of Agriculture (USDA). (1997). *Agricultural Resources and Environmental Indicators, 1996–97*. Agricultural Handbook No. 712, Natural Resources and Environment Division, Economic Research Service, Washington, DC.

United States Department of Agriculture (USDA). (2002). *The 2002 Farm Bill: Provisions and Economic Implications*. www.ers.usda.gov/Features/FarmBill/.

United States Department of Agriculture (USDA). (2003a). *Conservation Expenditure Data*. http://www.ers.usda.gov/Briefing/ConservationAndEnvironment/Data/table6_1_2 updated.xls.

United States Department of Agriculture (USDA). (2003b). *Emphasis Shifts in U.S. Agri-Environmental Policy. Amber Waves*. November 2003. http://www.ers.usda.gov/Amberwaves/November03/Features/emphasis_shifts.htm.

United States Department of Agriculture (USDA). (2006a). *Agricultural Resources and Environmental Indicators*, 2006 edn, ed. K. Wiebe and N. Gollehon. Economic Information Bulletin No. (EIB-16), July 2006.

United States Department of Agriculture (USDA). (2006b). www.ams.usda.gov/nop/NOP/standards/FullRegTextOnly.html.

United States Department of Agriculture (USDA). (2006c). *Renewable Energy and Bio-based Products*. Farm Bill Comment Summary and Background. http://www.usda.gov/documents/Renewable_Energy_and_Biobased_Products.doc.

United States Geological Service (USGS). (2002). *Hypoxia in the Gulf of Mexico*. website – http://toxics.usgs.gov/hypoxia/hypoxic_zone.html.

Urban, D. L. (2005). Modeling ecological processes across scales. *Ecology*, **86**, 1996–2006.

Vandyke, K. A., Kazmer, D. J. and Lockwood, J. A. (2004). Genetic structure of the alpine grasshopper, *Melanoplus alpinus* (Orthoptera : Acrididae). *Annals of the Entomological Society of America*, **97**, 276–85.

Van Valen, L. (1973). A new evolutionary law. *Evolutionary Theory*, **1**, 1–30.

Vera, F. W. M. (2000). *Grazing Ecology and Forest History*. Wallingford: CABI publishing.

Vickery, J. A., Tallowin, J. R., Feber, R. E. *et al.* (2001). The management of lowland neutral grasslands in Britain: effects of agricultural practices on birds and their food resources. *Journal of Applied Ecology*, **38**, 647–64.

Warren, J. (1995). Set-aside your weedy prejudices. *New Scientist*, **148** (2002), 48.

Warren, J. (2003). Isolation by distance in the crustose lichens *Candelariella vitellina* and *Placynthium nigrum* colonising gravestones in north-east Scotland. *Biodiversity and Conservation*, **12**, 217–24.

Warren, J. M. and Topping, C. J. (1999). A space occupancy model for the vegetation succession that occurs on set-aside. *Agriculture Ecosystems and Environment*, **72**, 119–29.

Warren, J. M., Raybould, A. F., Ball, T., Gray, A. J. and Hayward, M. D. (1998). Genetic structure in the perennial grasses *Lolium perenne* and *Agrostis curtisii*. *Heredity*, **81**, 556–62.

Watson, C. A., Kristensen, E. S. and Alroe, H. F. (2006). Research to support the development of organic food and farming. In *Organic Agriculture: A Global Perspective*, ed. P. Kristiansen, A. Taji and J. Reganold. Victoria: CSIRO Publishing, pp. 361–83.

Williams, B. and Warren, J. (2004). Effects of spatial distribution on the decomposition of sheep faeces in different vegetation types. *Agriculture Ecosystems and Environment*, **103**, 237–43.

Wilson, P. and King, M. (2003). *Arable Plants – a Field Guide*. Old Basing, Hampshire, UK: English Nature and Wildguides.

Woodcock, B. A., Lawson, C. S., Mann D. J. and McDonald, A. (2006). Grazing management during the re-creation of a species rich flood-plain meadow: effects on beetle and plant assemblages. *Agriculture, Ecosystems and Environment*, **116**, 225–34.

Working Document. (1998). *State of the Application of Regulation (EEC) No. 2078/92: Evaluation of Agri-Environment Programmes VI/7655/98*.

World Trade Organization (WTO). (2006) *Agricultural Negotiations: Background Fact Sheet – Domestic Support in Agriculture*. Webpage at www.wto.org/english/tratop_e/agric_e/agboxes_e.htm viewed January 18, 2006.

Wratten, S. D., Bowie, M. H., Hickman, J. M. *et al.* (2003). Field boundaries as barriers to movement of hover flies (Diptera: Syrphidae) in cultivated land. *Oecologia*, **34**, 605–11.

Wu, J. and Boggess, W. G. (1999). The optimal allocation of conservation funds. *Journal of Environmental Economics and Management*, **38**, 302–21.

Yachi, S. and Loreau, M. (1999). Biodiversity and ecosystem productivity in a fluctuating environment: The insurance hypothesis. *Proceedings of the National Academy of Sciences of the United States of America*, **96**, 1463–8.

Index

Printed in the United States
by Baker & Taylor Publisher Services